PRINCIPLES OF
EXPERIMENTATION
AND MEASUREMENT

PRINCIPLES OF EXPERIMENTATION AND MEASUREMENT

GORDON M. BRAGG

Department of Mechanical Engineering
University of Waterloo

PRENTICE-HALL, INC.
Englewood Cliffs, New Jersey

Library of Congress Cataloging in Publication Data

BRAGG, GORDON M
    Principles of experimentation and measurement.

    Includes bibliographies.
    1. Mensuration. 2. Errors, Theory of. 3. Proba-
bilities. 4. Experimental design. I. Title.
T50.B66   507'.2   73-19518
ISBN 0-13-701169-5

10 9 8 7 6 5 4 3 2 1

Printed in the United States of America

PRENTICE-HALL INTERNATIONAL, INC., *London*
PRENTICE-HALL OF AUSTRALIA, PTY. LTD., *Sydney*
PRENTICE-HALL OF JAPAN, INC., *Tokyo*
PRENTICE-HALL OF CANADA, LTD., *Toronto*
PRENTICE-HALL OF INDIA PRIVATE LTD., *New Delhi*

To Doris

Table of Contents

Preface

This book was written as an introduction to the process of measurement and experimentation. It is intended mainly for engineers, but the draft material has been used by students in both the physical and social sciences as an introduction to the field. The arrangement of material is designed to be useful to the student as he actually performs an experiment. Thus, the planning of an experiment is discussed in Chapter 2, the actual performance in Chapter 3, the treatment of results in Chapters 4 and 5, and the reporting in Chapter 6.

At my university the material of this book has been used for several years in a one-semester lecture and project course. The project has been emphasized in this course and this accounts for the small emphasis placed on exercises in this text (except for the statistics section). A series of example projects attempted by students is listed in Appendix III. The contents of this book have been determined in large measure by the students' need to know the material in order to properly complete their projects.

An attempt has been made to make the material reasonably self-sufficient in order that a student working on his own may be able to find the book of use.

The first acknowledgement must go to my students over the past few years, the answers to whose questions form the essence of this book. Thank you to Bruce Hutchinson whose notes formed the basis for parts of Chapter 4, and to Ben van der Hoff whose notes were the basis for Sections 5.7 and 6.11. My thanks also to Tim Topper and John Hanson whose comments have been so helpful and to Bobbie Taylor who typed the many drafts.

Gordon M. Bragg

PRINCIPLES OF
EXPERIMENTATION
AND MEASUREMENT

1

The Place of Experimentation
and Measurement in
Science and Technology

The performing of experiments and measurements in an efficient and useful way is the subject of this book. The methods described are quite general and are used in all fields of science and technical endeavor.

1.1 The Obtaining of Information

In the fields of scientific and engineering study and research there are two main ways to obtain information: analysis and experiment. Analysis is the use of agreed upon theories and mathematical formulas to predict and analyze physical situations. In experimentation we return to the actual phenomena under study and actually measure what happens.

The scientific and technical advances over the past two centuries have provided for us a huge quantity of theory and analysis. For this reason, much of undergraduate education in the sciences and technologies is devoted to presenting and explaining this analysis. However, in spite of its quantity, this analysis does not explain (at least in sufficient detail) many phenomena of interest to us. We must still have recourse to experiment and measurement even in well-understood subjects. It is the purpose of this book to introduce to you the professional practice of experimentation and measurement.

1.2 The Professional Approach to Measurement

One of the fundamental aspects of professional practice is that the problems are not well formulated. Consider law and medicine. Lawyers spend considerable proportions of their time on deciding the point of law which is in question for a particular case or client. The client is not in a position to do this himself. If he were, in many cases he could decide for himself whether his activities were within the law, whether he should prosecute, etc. In medicine, diagnosis is a fundamental aspect of the treatment. Once the diagnosis is made, medical texts make the course of treatment clear in many cases. The problem in engineering and the sciences is similar. Definition of the problem is a most important aspect of the profession.

Difficulties arise when these professional aspects are taught in an academic environment. The techniques required for doing this work are not *hard*. That is, numbers, mathematics, lists of procedures, etc., are not sufficient knowledge for a competent professional. The medical and legal professions have known this for centuries and have, respectively, developed the clinical method and case study methods of teaching these techniques. In both cases, examples are worked or treated under the guidance of an experienced professional. In experimentation the obvious analogy to these methods is the project method. This book is designed to aid you in accomplishing projects which involve measuring some quantity.

Why measurement? In the past decade "science" has become a major (and occasionally the only) component of undergraduate curricula in the professions. This has been especially true of engineering. This type of curriculum emphasizes hard information such as applied mathematics, physics, and chemistry as a necessary ingredient in an engineering curriculum. In the past few years, however, it has become clear that when carried to extremes, this emphasis trains scientists and not engineers. This book is an attempt to redress the balance in two ways. We shall attempt to present the *soft* side of the professional approach, and we shall apply it to measurement. It is considerably easier to obtain information analytically compared to experimentally if the analysis exists and is accurate. Therefore, much of engineering education consists of learning this analytical material. However, all analysis must be compared with the real world by tests. These tests involve measurements, which are another path to physical knowledge.

In summary then: *We shall study the professional approach to obtaining measurements with the objective of learning (1) how measurements are obtained in realistic situations, (2) how an organized professional approach can aid us in this as well as other problems, and (3) that measurement is a respectable way of obtaining physical knowledge comparable with analysis.*

In the past few years there has been considerable research effort devoted to quantifying and analyzing the problem-solving process. This work includes the studies of systems analysis, operations research, linear programming, statistical design and analysis of experiments, control theory, and several other fields. These fields are all comprehensive subjects in themselves. Part of our objective in this book will be to introduce some of these subjects very briefly and to outline a few of the things which these tools can accomplish.

1.3 The Scientific Method

Our factual knowledge of the physical world is based upon science. How is this knowledge obtained? The "classical" scientific method assumes three stages:

1. A *hypothesis* attempting to explain a phenomenon is proposed; for example, that force is proportional to the product of mass and acceleration. Force, mass, and acceleration are also defined. All consequences of the hypothesis must be logical.
2. An *experiment* or measurement is made in the environment where the hypothesis is assumed to be true. In the example, force, mass, and acceleration are measured over a range of each.
3. The experiment and the hypothesis are compared. If the hypothesis seems to be accurate and acceptable to the community of people most competent to judge both experiment and hypothesis, then the hypothesis becomes a *theory*.

This simple concept of the scientific method has been much battered about by historians, philosophers, and scientists, particularly in the last 20 years. The actual process of an investigation can be much different from this in many cases. For example, it is common for hypotheses to be proposed on the basis of experiments rather than vice versa. It is common to assume hypotheses which are known not to be strictly accurate in order to explain phenomena. It is also common for scientists to speak of *models* other than theories. A model can be a physical description of a phenomenon which is adaptable to mathematical analysis. The mathematical model then is proposed not so much as an explanation of nature but rather as something which it is hoped will behave in a way similar to the phenomena.

Our interest in this procedure stems from our involvement in the experimental part of the process. The pure scientist performs experiments to *discover* basic behavior and to *compare* his results with models and theories.

1.4 The Engineering Approach to Measurement

The essential difference between engineering and science may be described by saying that the objectives of "pure" engineering are creative while those of "pure" science are analytical. Both require information about the physical environment and also, in modern terms, about the social and economic environment. Because of this, the measurement and analysis techniques are common to both, with engineering more dependent on science than vice versa. In addition, the detailed application of measurements is rather broader in the engineering case. Consider the following examples of measurements which would have little usefulness or relevance to scientists:

1. Measurements of factory production rates on various days.
2. Measurements of radio reception at various distances from the aerial.
3. Measurements of movement of house foundations due to frost.
4. Measurements of variations of dimensions in manufactured products due to machine tool wear.

It is obvious that the objectives of engineering measurement are intimately tied to the objectives of the engineering project. This importance of defining the objective will be referred to again in Chapter 2. It can be seen that while we shall use the information and the tools of science, we shall use them to quite different ends.

1.5 The Variability of Measurements

Measurements are inherently variable. No measurement can be repeated exactly. In many simple measurements, such as measuring the dimensions of a room, the precision is not sufficient to detect these variations. In other measurements, the inherent variability is obvious. The daily production of a large factory will vary from day to day due to hundreds of different factors. Repeated precise measurements of the same object with the same instrument will produce variations. The reading of a micrometer will vary, depending on such quantities as

1. Operator changes.
2. Room temperature.
3. Dirt on the workpiece measured.
4. Tension on the spindle.
5. Zeroing error.
6. Misreading of the dial.
7. Interpolation errors.

This variability is fundamental to all measurement systems.

In atomic physics, this variability has a theoretical basis. Heisenberg's uncertainty principle states that "The product of uncertainty in position and uncertainty in momentum may not be less than a constant."* This means that the more accurately we know the position of a small particle, the less accurately we know the momentum of that particle and vice versa. Intuitively, this may be explained by the fact that if we wish to know the position of a particle, it is necessary to slow it down, thus changing the momentum. If we wish to know the momentum, we cannot slow it down, and thus we are uncertain about the particle position. To simply observe the particle will not let us out of this impasse since observation implies light, and light involves photon bombardment of the particle.

The problem in macroscopic measurements arises from a different source and is not so generally based on Heisenberg's principle. The macroscopic problem arises from two facts:

*The constant is $h/4\pi$, where h is Planck's constant.

1. It is impossible to measure something without using the phenomenon to influence the measuring instrument. This implies that the measurement changes the phenomenon (similar to Heisenberg's principle).
2. It is impossible to remove totally the influence of variables which influence the process but which are not wanted in the measurement.

An example of the first would be the placing of a large thermometer into a cup of coffee. The thermometer will cool the coffee. The second problem occurs because the coffee is cooling down in time, because the coffee has convection currents in it which cause and are the result of temperature differences, because there is a finite error in the calibration of the thermometer, etc.

In simple experiments the variability is often smaller than any necessary accuracy required. In difficult measurements it is possible for variations in results to be many times larger than the measurement required. In astronomical radio telescopes it is not uncommon to receive signals of importance which are many times less strong than the associated "static."

1.6 Exercises

1.1. Describe how the measuring process affects the thing measured in the following cases:
  (a) Measurement of light level in a room with a light meter.
  (b) Measurement of the thickness of a human hair with a micrometer.
  (c) Measurement of an assembly line worker's output by an observer.
  (d) Determination of tooth decay by X-ray.
  (e) Determination of the breaking strength of glass by bending glass rods.

1.2 Report on the process of discovery of any of the following:
  (a) Newton's laws.
  (b) The transistor.
  (c) The special theory of relativity.
  (d) The laser.
  (e) X-rays.

1.7 Suggestions for Further Reading

Arons, A. B., and Bork, A. M., *Science and Ideas, Selected Readings,* Prentice-Hall, Inc., Englewood Cliffs, N. J., 1969.

Hanson, N. R., *Patterns of Discovery,* Cambridge University Press, New York, 1965.

Koestler, A., *The Sleepwalkers,* Hutchinson & Co. (Publishers) Ltd., London, 1959.

Wightman, W. P., *The Growth of Scientific Ideas,* Yale University Press, New Haven, Conn., 1953.

Wilson, E. B., *An Introduction to Scientific Research*, McGraw-Hill Book Company, New York, 1952.

2

Defining the Problem

The most important aspect of any measurement or experiment is the problem definition, because a proper definition will indirectly include the complete statement of the process to be followed. Four examples of this are discussed below.

2.1 Examples of Problem Definition

Example 1

It is required to find the height of waves in Lake Huron. The wave heights in a situation such as this range from less than 1/10 in. to several feet. A measuring instrument precise enough to measure wave heights of 1/10 in. (say to 10%) would require a measurement precision of 1/100 in. An instrument of this sophistication is unlikely to be strong enough to resist waves several feet high. The large waves will occur only during storms and probably will be highest in the winter, a period when it will be very difficult to service the instruments and record from them. In addition the large waves will be very infrequent. How long a sample is required and under what range of conditions? Obviously a measurement process which enables all these conditions to be measured would be extremely expensive.

If we consider the reason for the measurements, we can eliminate those which are not required. If the wave heights are required for ship design, then the frequency and magnitude of the largest waves will be important. In fact, in the extreme case a new measurement may not be required at all. A search of the log books of ships in Lake Huron during a very bad storm sometime in the past 10 or 20 years might well turn up an estimate by the ship's captain of the highest wave he had seen on the lake in his experience. This could well be sufficient. Modern commercial ships are seldom less than 200 ft long. Any wave less than, say, 2 ft in height is unlikely to have any significant effect on these ships. We might therefore consider a study applying only to waves higher than 2 ft. In addition, Lake Huron is closed to shipping for approximately 3 to 4 months each year. It would not be necessary to measure during this period.

If, now, our purpose is to study the effect of wind on creating possible currents in the lake, we have a quite different situation. In this case the action of wind on quite small surface waves may be a mechanism for transferring energy from the wind to the water. Without measurements (and some quite sophisticated theory) we cannot assume that the small waves are not important or for that matter that large waves (being infrequent) are not important either. Our approach must be quite different in this case.

If our purpose were to determine the *shape* of waves, we would certainly expect this to vary with wave size. However, here we may need only to measure a few waves in each height range (say 50). The relative frequency may well be unimportant.

Each of these cases will require a quite different approach to the basic problem.

Example 2

It is required to find the "best" brand of C cell batteries. The problems which would arise in obtaining a definitive answer to this question include

1. The batteries' decay in effectiveness if left on store shelves.
2. Batteries with large initial voltages may or may not have large voltage decay rates.
3. The voltage of a "new" battery varies among batteries.
4. Batteries of the same brand may vary in quality as widely as the variation among brands.
5. The batteries' quality may depend on whether or not they are used continuously.

If specific reasons for the measurement of these batteries are defined beforehand, then some of the problems may be resolved.

For certain types of modern electronic equipment there is a voltage below which the circuit is totally inoperative. If the batteries are to be used in a circuit of this type, then the major requirement is that they exceed this critical voltage. If the batteries are to be used infrequently, it is possible that the tests made should allow for periods of nonoperation of the batteries.

If a system works better with higher voltages, then the beginning voltage may be a suitable testing parameter. This would be particularly true if the batteries were frequently replaced.

If the test is being carried out for an infrequent user of batteries, the question may not revolve around brands at all. The basic question would be: Where should a buyer purchase batteries in order that their time on the shelf will have been short?

This example is treated in more detail in Chapter 5.

Example 3

It is required to determine voltage variations in domestic power outlets. This voltage can vary quite significantly. The variation can be due to the following:

1. Large or small loads on the whole power system. Maximum loads occur in the morning around 9 A.M. when factories start up and around 5 P.M. when factories are still operating and home use for cooking is also large. Over a larger span, the peak power use in most systems is at approximately 5 P.M on one of the few days before Christmas. The uses then would include Christmas lighting, house lighting in the early evening, large meals being cooked in homes, and peak heating requirements as well as industries which are still operating.
2. On each subsystem the load may influence the voltage. It is common to notice lights flickering when a large motor is started up in a house.
3. Lightning and flooding can cause shorting and power surges.
4. Other large power users can be on the same circuit.
5. Controls on maximum voltage variation allowed may change from area to area.

In a situation where any or all of these variations can occur, we must consider the reasons for wishing to know the local voltage.

The specific requirements for the measurement could be to

1. Determine if a piece of electrical equipment may be used. Certain types of electrical equipment can be damaged or fail to work if the voltage is too high or too low. The equipment may only be in use during the 9 A.M.–5 P.M. period.
2. Determine if the power authority is operating within its own standards.
3. Determine if the power authority is working outside its own standards consciously. This occurs from time to time when demand is high. Several areas of the eastern United States have had periods of brown out in the last few years.

Example 4

This example is drawn from the personal experience of Sir Geoffrey I. Taylor, a man who has made many important contributions to the field of mechanics.

I had studied fluid dynamics as a student and was applying my knowledge to Meteorology when the 1914–18 war started on August 4th. Like most of my friends I volunteered for service in the army and suggested that I might set up a weather forecasting unit in the field. The officer to whom I made this proposal did not seem to doubt that I could tell what the weather was going to be—as he might very well have done—but thought the knowledge would be of no value to the army in the field. "Soldiers don't go into battle under umbrellas; they go whether it is raining or not" was his comment.*

These examples are intended to illustrate the necessity to carefully define the problem, since the resulting tests will be highly dependent on this.

It is worth mentioning here that an engineer will seldom, if ever, receive a problem which has been properly defined. As with other professions, this is considered to be one of his fundamental duties. The medical profession considers diagnosis to be an extremely important aspect of the practice of medicine (possibly even the most important). The law profession spends considerable time identifying the point of law in question on any particular case. It is worth mentioning again that in medicine the method of teaching diagnosis revolves around the clinical method, and that in law the teaching revolves around the case method. The engineering profession equivalent is the project method.

2.2 Determining an Appropriate Problem Definition

One frequently hears the phrase "Now I know what the problem is." We must now consider this most important aspect of most scientific and technical work.

No simple rules as to how to define a problem can be laid down. One point, however, is quite clear. If the problem definition does not suggest the method of proceeding (consider again medical diagnosis and case law study), then it is an improper definition. Further, the definition should include within it a set of statements or ideas formed in such a way that we can tell when the job has been accomplished. This obvious statement has a number of implications with which we shall deal in Section 2.3.

Measurement experimentation can easily be considered as a special case of design in the sense that there is a creative or innovative act of designing a measuring process. The field of systems analysis also gives guidance concerning measurment problems. Systems analysis may be defined as the study of relationships between objects. Without going into the philosophy or methodology of these subjects we wish simply to point out that in both disciplines a major emphasis is placed upon definition of the problem. Much of the material in this book is formulated upon the basis of using a systems or design approach to measurement.

It is not necessary to fix in written form an irrevocable set of statements which must be followed as a problem statement or definition. This is particularly true in a measurement situation. Briefly, if you know exactly what you are doing in an experiment, there is no sense doing it.

*Address to the Third Canadian Congress of Applied Mechanics, Calgary, Canada, May 1971. Quoted with permission.

A few years ago in the setting up of an undergraduate experiment, a set of 10% resistors was purchased. Their resistance was to be measured by the undergraduates to illustrate the Gaussian distribution (see Section 4.5). The result was to look like Figure 2.1. The problem definition for the instructor was then to

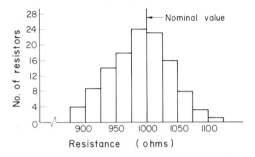

Fig. 2.1. Proposed result of resistor tests.

demonstrate that the resistors purchased did indeed conform to this type of distribution. In actually testing the resistors, a distribution was found which looked like Figure 2.2. The lack of resistors near the nominal value showed that

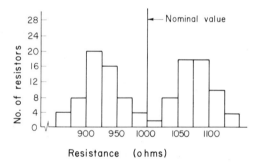

Fig. 2.2. Actual result of resistor tests.

the basic assumption was incorrect, because in the manufacture of resistors, the process is rather poorly controlled. 5% resistors are commercially available and are "manufactured" by testing each resistor and labeling those closest to the nominal value as 5% resistors. This means that 10% resistors (the leftovers) contain few close to the nominal value. In the end the experiment was "cooked" by making available to the students a very carefully selected set of resistors which gave the distribution of Figure 2.1. This point may serve as an illustration of why the author is not in favor of "carefully designed" laboratory work to illustrate basic principles. More importantly for our purposes, it shows why there may be perfectly good reasons for changing the objectives and definition of a program during the actual work. Proper study of the methods of production of resistors would, of course, have enabled the project to be done properly in the first place.

We shall now consider four aspects of problem definition which are particularly important in measurement: setting criteria by which the end result may be judged, planning of the experimental program, aspects of problems which have many variables, and searching for measurements or information made available by others.

2.3 Setting Criteria

A few years ago *Scientific American* magazine sponsored a paper airplane contest where prizes were to be given for beauty, time in the air, distance, etc. The planes were to be made by the contestants and flown by a group of graduate students who would give each plane repeated flights. There was considerable difficulty in judging among planes. Some hit the roof of the auditorium. Some were difficult to launch in a specific direction. Should a curved flight be measured along its curve for the distance event? The difficulty of judging, of course, was based upon the criteria applied. Part of the pleasure of paper planes is in the unexpected things they do. This could not be judged directly. Without attempting to belittle the contest (which was obviously not meant to be taken too seriously), the judges encountered difficulties because criteria for actual airplanes were not directly applicable to the paper ones. A number of undergraduate projects on paper airplanes at the author's university have in past years been inconclusive due to the difficulty in setting up proper performance criteria.

The setting of standards of accuracy, number of measurements, type of instrumentation, manner of presentation, etc., are all implied if the measurement is well planned at the definition stage. An obvious example is the set of specifications available on high-fidelity sound equipment as supplied from the manufacturer's measurements. Here some specifications on the equipment and the precision and accuracy of the measurements are obviously well beyond the capability of the human ear. These measurements have been made, and the equipment designed, for advertising purposes.

The breaking strength of eggs is a favorite undergraduate project and has even been the subject of postgraduate study at at least one university. If posed as an engineering problem, we may wish to reproduce as accurately as possible the circumstances where the breakage occurs. If breakage in packages is the problem under study, our experimental test is not obvious. Simple crushing tests and dropping tests of the uncovered egg will produce numbers (force and height), but their relation to the original problem is not clear. A more appropriate test might be to test breakage while in the packages. Here again, dropping and crushing could be considered; however, the direction of the forces and the orientation of the packages on landing will now be important. A proper experiment will strongly depend on the way in which the actual problem has been defined.

2.4 Planning

Part of the original definition of the problem will include a plan of the actual performance of the test. No experiment should proceed without a full and detailed plan including the number of measurements to be taken and the way in which they will be reduced, plotted, etc. In simple problems this aspect may not be important. However, improper planning can make an experiment worthless. In a project performed by a student a number of years ago, the air leakage through a window frame was to be measured. The flow through the window was to be obtained by means of a flowmeter. After installing the meter and running the apparatus, it was found that the maximum air flow was too small to have any effect on the flowmeter chosen. Errors of this type are particularly easy to make.

A study of the average height of young males was undertaken, and the heights of 120 engineering students were recorded. The average height was then produced as the average height of young males. The problem with this conclusion lies in the sample taken. Other research has shown that university students are not true cross sections of the whole population. Attendance at university depends on economic background, I.Q., etc. If any of these other aspects are related to height, then the sample has been wrongly chosen. The problem of correct sampling procedures will be considered further in Chapter 4.

The choice of variables is the major step in planning an experiment and can be very difficult. As an example, the study of fine dusts in the air has been the subject of numerous studies with applications to air pollution control and industrial hygiene. The majority of these studies has concentrated on such topics as the size and shape of the particles and the velocities at which they fall. In the last few years it has become apparent that the electrostatic charge on these dust particles can completely determine what happens to a particle. In turn the electrostatic charge depends on the size, shape, and material of the particle. The relative humidity also has a large effect on the particle charge. With this information (which does not seem strange or unusual in retrospect) the planning of further research on dusts is much changed from a few years ago.

In many cases the variables are already determined either from the theoretical knowledge or from previous experiments. The values of the variables, however, are required in a new situation. If a camera is being used to photograph stars, the lens opening and exposure time are required. This is a difficult form of photography, but the principles of operation are exactly the same as in the simplest snapshots.

2.5 Problems with Many Variables

Probably the most obvious difficulty in real measurement problems revolves around the fact that what is to be measured depends on many variables. The flow of traffic through an intersection, for example, depends on

1. Direction considered.
2. Time of day.
3. Time of year.
4. Weather.
5. Road repairs.
6. Accidents and tie-ups.
7. Method of counting. (Is a bicycle a vehicle?)
8. Timing of traffic lights.
9. Speed of traffic.
10. Arrangement of traffic signs (e.g., four-way stop or two-way stop) and many others.

Several possible approaches are available, depending on what is required for the study.

1. It may be neither possible nor desirable to control these variables. A proper sample including them as they occur is required. The traffic flow case might be an example of this. A study over several complete workdays may be appropriate.

2. Only extreme cases need be considered. Here the extreme limit or a value close to it is required. Again, traffic flow can be an example. If the maximum traffic flow is important, then a sample taken from 4 to 6 P.M. on Friday evenings may be appropriate. For both cases 1 and 2, statistical reasoning is extremely important in order to make sure that the sample represents the quantity required. It is common practice in the design of hydraulic structures (dams, spillways, etc.) to design for conditions which are exceeded only once in 50 to 100 years. It is this condition which must then be obtained.

3. If one phenomenon predominates an event, then we may wish to study this phenomenon in complete isolation in order to understand it further. This represents the classical scientific experiment where we "vary one thing and keep all others constant." As an example, consider the time constant of a thermometer. It is a fairly well-known fact that a thermometer which undergoes a nearly instantaneous temperature change will take a certain finite amount of time to approach the new temperature closely. Theoretically, it takes an infinite time to completely achieve the new temperature of its surroundings. The time for the thermometer to record a given percentage of the total temperature difference is constant. The percentage chosen is 63.2% (for analytical reasons), and the time is called the time constant (see Figure 2.3). If we wish to test these statements and to obtain the time constant, we need only to reproduce the conditions of the statement as closely as possible:

a. Instantaneous temperature change.
b. Same thermometer for each test (the time constant will, of course, change for different thermometers).
c. Vary the temperature difference.
d. Check the constancy of the time constant.

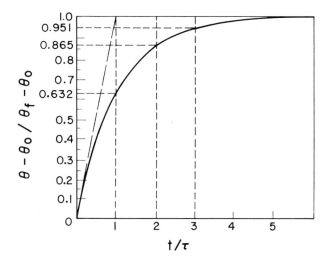

Fig. 2.3. Time taken for a thermometer to reach its final temperature. The horizontal axis is time divided by the time to reach 63.2% of its final value. The vertical axis is the temperature at that time minus the original temperature over the final temperature minus the original temperature.

Other variables such as a time-varying temperature at the new level have been excluded.

4. If a quantity is desired which depends on several other quantities and its dependence is required, then the problem can be a quite difficult one. For example, the classical problem of the motion of three bodies acting by gravitation on each other is not calculable in an exact sense. Two techniques are helpful in studies of this type. Dimensional analysis enables variables to be grouped together for purposes of study and can be a very powerful tool. Dimensional analysis is treated in Section 5.5 and is frequently useful in the planning stage of a problem. The second approach is more difficult; to revert to a type 3 problem and vary one parameter at a time. The difficulty of dealing with this is illustrated by considering that a function of one other variable is described by a table. If a variable is a function of two others, then a book of tables is required. If of three others, then a shelf of tables, and if of four others then a small library would be required. It is not uncommon to find functions of 30 and 40 variables. Local weather prediction is an excellent example. In practical cases the range of variation of the variable is often small and the tabulation is made. Large proportions of engineering and scientific handbooks (of which there are many) are devoted to tabulations of this type within the usual ranges. It should be noted that extrapolation of this information beyond the tabulated range is most dangerous. Extrapolation is considered in Section 5.8. The problem of correct sampling procedures will be considered further in Chapter 4.

2.6 Searching the Literature

The modern information explosion has made available to engineers and scientists an extremely large body of information and knowledge for use in solving particular problems. In fact, this body of information is too large for convenient use. The problem has in many ways become finding knowledge other than production of knowledge. One writer, on finding that the rate of scientific publications is increasing exponentially, determined that if the rate were maintained, the weight of scientific publications would equal the weight of the earth by the year 2200. Appendix III contains a list of over 100 student measurement projects. It is unlikely that any of these projects lacks less than a dozen suitable and relevant references on the subject or on subjects so closely allied that the information is still relevant. The problem is to find the material at as early a stage as possible in the project. The objectives of the search will be to

1. Find out if the planned work has already been done.
2. Understand the physical, economic, social, etc., processes operating on the problem, that is, determine the variables.
3. Determine the best method of proceeding and the pitfalls of other work.
4. Determine where information is lacking.
5. Put your own work in proper perspective.
6. Determine legal conditions surrounding the work (for example, boiler pressures or pollution levels).

Sources begin, of course, in a library. While other information methods are growing quickly, the written word still contains the vast majority of known factual knowledge. Every student should have some familiarity with the reference and bibliographical resources of a modern library. However, knowledge obtained only through professional librarians is not sufficient since frequently only the person working on a project is a competent judge of the worth of reference material. Judging other work becomes a necessary part of the accomplishment of a project.

There are four other recommended beginning points to a reference search. The beginning is most important since it is usual for one reference to lead to another. These four sources are texts, handbooks, abstracts, and specialists in the field. Textbooks cover specific fields broadly and refer back to the more detailed studies. Handbooks contain summaries of information usually aimed at the specialists in a field. They are difficult to learn from and often lack references. In technological fields they can become outdated quickly. A large amount of technical information, and almost all the newest information in technical and scientific fields, lies in journals and magazines. It is the object of abstract journals to summarize this information on a regular basis. Examples are *The Engineering Index, Applied Mechanics Reviews, Chemical Abstracts,* and *University Microfilms.*

Undoubtedly, the best beginning to a literature search is to consult a specialist who has already studied the subject and has a broad knowledge of the best sources, which enables him to give a sense of proportion to the information and guidance as to quality, relevance, etc. It is not uncommon, however, that the performance of a project makes the person who does it the only available specialist on that very narrow subject.

2.7 Example

The points made in this chapter may be illustrated by an example from the field of industrial hygiene.

Asbestos dust is toxic and can cause several diseases including a degenerative disease of the lungs called asbestosis. For this reason in most countries of the world a maximum allowable number of particles per unit volume of air has been established by law. The objective of several studies has been to determine the best method of measuring these particles.

An appropriate problem definition for this type of study depends on the accuracy, cost, and consistency of the method. It also depends on simplicity since relatively unskilled operators must be trained to use the method. The method must be accurate enough to enable offenses to be determined fairly and consistently.

Several problems arise. The dust of asbestos is long and thin and in real situations occurs with other dusts, such as wool, which are not harmful. If particles are to be counted, there is a size below which modern instrumentation cannot measure. In addition, medical studies seem to indicate that the larger particles of asbestos are more dangerous to health.

The problem definition in common use today is based upon these points. It could be briefly put as follows: The objective is to measure at least 200 particles and the volume of air in which they occur. Only particles with a length to width ratio of greater than 3 to 1 will be counted and only those longer than 5 microns. A micron (μ) is 10^{-6} meters. A criterion for successful measurement is that the method should give consistent (although not necessarily accurate) results if repeated. The statistics of Chapter 4 will enable this criterion to be stated quantitatively.

The actual test used in this case consists of sucking air onto a fine filter which collects the dust. A piece of the filter is then placed on a microscope slide and dissolved chemically, leaving the asbestos; then the slide is viewed under a 400-power microscope. The additional details required to assure consistent results consist of approximately 10 pages of instructions.

This involved and complicated experimental plan is necessary in this case because of the social implications and legal requirements of a problem which has many variables interfering with the one of interest. The legal requirements for the above test are presently that no more than two fibers per c.c. longer than 5 μ and of a length to width ratio of greater than 3 to 1 should be present in any environment where people are working. This standard is common in Britain, the United States, and Canada, and similar standards are in use elsewhere.

2.8 Exercises

2.1. Discuss the proper formulation of the following problems:
   (a) Find the total area of leaf present on a mature maple tree in order to determine the amount of sunlight which the tree can absorb.
   (b) A cast iron table surface has been manufactured as a work surface for delicate dimensional measurements. The objective is to determine (1) the smoothness and (2) the flatness of the surface.
   (c) Dust as seen through a microscope must be sized. The particles are of many different shapes, from long thin fibers to near-spherical particles. The dust has been taken from the air with the objective of determining the degree of air pollution.
   (d) In a study of car safety, the time required for a person to sense an emergency and move his foot to the car brake must be determined.
   (e) The weight of snow is required for the design of roofs, snowplows, etc.

2.2. To familiarize yourself with your technical library, look up the following material (if available in your library) and take a few moments to glance through them:
   (a) Eshbach, Ovid W., *Handbook of Engineering Fundamentals,* John Wiley & Sons, Inc., New York, 1952 (handbook).
   (b) Lion, Kurt S., *Instrumentation in Scientific Research*, McGraw-Hill Book Company, New York, 1959 (monograph).
   (c) *American Society of Mechanical Engineers Boiler Code* (code).
   (d) *Engineering Index* (abstracting journal).
   (e) *Journal of Scientific Instruments* (journal).
   (f) Doebelin, Ernest O., *Measurement Systems*, McGraw-Hill Book Company, New York, 1966 (textbook).

2.3. Restate the following general studies in such a way that the numerical objectives of the study are obvious. Narrower objectives may be assumed; e.g., "the accuracy of automobile speedometers" may be more specifically narrowed to "the percentage error of speedometers as a function of velocity."
   (a) Methods of determining the value of gravitational acceleration.
   (b) The strength of drinking straws.
   (c) The growth of a town or city.
   (d) The acceleration of cars.
   (e) Temperature distribution in a house.

2.4. Propose a system for measuring the loads on trucks which use a given highway. For reasons of economy every truck cannot be measured. Remember that a decision must be made as to what trucks are overloaded and that the weighing serves a law-enforcement function. State objectives and suggest uses for the resulting data.

2.9 Suggestions for Further Reading

Hall, Arthur D., *A Methodology for Systems Engineering,* Van Nostrand Reinhold Company, New York, 1962.

Edel, Jr., D. Henry, *Introduction to Creative Design*, Prentice-Hall, Inc., Englewood Cliffs, N. J., 1967.

Roe, P. H., Soulis, G. N., and Handa, V. K., *The Discipline of Design,* University of Waterloo Press, Waterloo, Ontario, 1969.

3

The Mechanics of Measurement

In this chapter we shall attempt to outline the mechanism of the actual measurement process. There are many books devoted to this particular aspect of the complete process; a number of them are listed at the end of the chapter. The purpose of this chapter is to illustrate the large range of *types* of measurement and the even larger number of actual instruments and methods of using them. The general considerations which control the sophistication and accuracy of the instruments are briefly outlined. We shall make no attempt to describe actual instruments except as examples of general principles.

3.1 Types of Physical Measurement

For a physical measurement to be made, we must produce a response of some sort from the thing we are trying to measure. That is, the measurand (the thing measured) must produce a response on an instrument of some sort even if the response is as insignificant as the reflection of light in such a way that we can measure its length with a rule. There are two implications. First, we cannot measure a phenomenon without transferring energy from it. Second, this will change the thing we measure, as in the case of a large thermometer placed in a small cup of coffee. The presence of the thermometer will cool the coffee. This, in turn, will influence the resulting temperature measurement. Since all measuring processes use energy, we can classify them according to the type of energy which they use (thermal, optical, electrical, etc.). Many processes, of course, use several types.

The *signal* or *information* from the measuring instrument can also take many forms. In terms of electrical terminology the signal can be analog or digital. Analog information is continuous information. A dial meter reading, a sound level, a length on a ruler, a tracing by a pen on a chart, and a light intensity are all examples of analog information. In theory, analog information can be varied in infinitely small quantities and is of the form of a function. Digital information, however, is of the yes-no, on-off, black-white, or integer form. The number of students in a class, the pages in a book, and the record of whether a switch is on or off are all digitized information.

Information can be carried on waves. For example, a sine wave can give information through its amplitude, frequency, or phase. The system can be pulsed, and the pulses can bring the information through their amplitude, frequency, position, width, or presence or absence in a train. The most common types of measurement, such as a meter reading and length or temperature measurements with a ruler or thermometer, may then be called in this terminology *analog direct current*.

These various types of signals each have certain abilities which are more appropriate to different measurement situations. For example, digital information is often more precise than analog information (but not necessarily more accurate). However, analog information in some situations is better for fast response to time-varying quantities. Decisions concerning the best form of information transport can be quite difficult in complicated systems such as chemical process control and spacecraft metering.

The instrument used in the measurement transfers energy. An energy transfer device is called a *transducer*. Transducers can have external sources of energy (oscilloscopes) or be driven by the phenomenon (thermometers). They can have a number of elements (computers coupled to the instrument) or only one (a ruler).

Instruments can be balanced or null reading as in a mechanical or Wheatstone balance, or they can be unbalanced as in a meter or manometer. Typically, balanced systems are more accurate than unbalanced systems and are often used to calibrate unbalanced systems.

The measurement itself may be constant or time varying. If the system is time varying, the ability of the instrument to respond to this variation accurately becomes important.

The measurand may be nonrepeating such as an explosion, an eclipse, or an earthquake. In this case difficulties arise since the experiment may not be repeated in time and may require repeating by duplication of equipment.

The range of sensing and measuring instruments is obviously wide even in general terms. When we deal with particular applications the number becomes astronomical. The general theory of instruments is quite general, however, and encompasses all types. The actual discussions of energy and its transfer, of course, depend on basic physical principles. The design and use of instruments as a system, however, are the province of control theory, systems analysis, and, in some cases, operations research. These subjects owe a great deal to the impact of electrical technology on modern engineering. However, the basic methods of analysis are easily and properly used in all types of measurement systems.

3.2 Transducers

All energy transfer devices have been defined as transducers, and hence all measurement devices are transducers. The actual design of physical transducers depends on the basic laws of physics and chemistry. We shall begin our study of them by outlining the tremendous range available. We shall follow a presentation similar to Stein.[*]

A *passive* transducer has two terminals (input and output).

An *active* transducer has three terminals (major input, minor input, and output). Each of the inputs or outputs can take the form of various types of energy, for example, chemical, optical, mechanical, electrical, thermal, magnetic acoustic, and nuclear.

Examples are given in Tables 3.1 and 3.2. Each input and output transfers both energy and the required *signal*, that is, the information of the measurement.

Table 3.1.   Passive Transducers—Examples

Input / Output	Chemical	Optical	Mechanical	Electrical	Thermal	Magnetic	Acoustic	Nuclear
Chemical		Camera film						
Optical	Litmus paper	Microscope		Galvanometer	Pyrometer			Cloud chamber
Mechanical			Spring balance		Thermometer	Compass		
Electrical			Phonograph cartridge		Thermocouple			
Thermal								
Magnetic								
Acoustic			Tuning fork	Earphones				
Nuclear								

Table 3.2.   Active Transducers—Examples

Example	Major Input	Minor Input	Output
Oscilloscope	Electrical signal	Electric power	Optical screen
Strain gage	Mechanical strain	Electric power	Electric signal
Thermostat	Temperature	Voltage	Electric current
Pressure transducer	Pressure	Voltage	Current
Radio	Electromagnetic radiation	Electric power	Acoustic radiation
Microphone	Acoustic radiation	Electric power	Electric signal
Pan balance	Weight of unknown	Known weights	Balance reading

[*]P. K. Stein, *Measurement Engineering*, published by the author (1969).

In an active transducer the major input carries the information. The minor input represents an additional method of *controlling* the information. For example, the information and energy may be changed in type, amplified, or controlled as to frequency. The minor input can be used to discriminate between two or more ranges of information, to filter out noise or static, etc. A radio is an example of all of these types of control by the minor input. Since changes occur both to the signal and the energy in both active and passive transducers, it is apparent that these changes may not all be to our advantage. The active transducer with its possibility of control then represents in theory a "better" instrument. However, this is not necessarily true since the active transducer in effect represents more *hardware* between the measurand and ourselves. Each piece of hardware represents more possibility of error and of expense.

Every transducer is made up of elements or parts, and these elements can treat energy in one of three ways:

1. They can store potential energy (springs, capacitor, pressure chambers, etc.).
2. They can store kinetic energy (inductance, flywheel, etc.).
3. They can dissipate power (resistors, dynamometers, dampers, etc.).

Using the transducer terminology, as we have, it is easily seen that the range of *types* of transducer is very large if the types are based upon energy considerations. If the information type is also considered (as in Section 3.1), we have a very large range indeed. You should also note that the concept of a transducer is artificial in the sense that the definition depends on where we set the limits of our *system*. A radio is a transducer, but so is the aerial within it, as is the speaker or the amplifier. The definition of the transducer will depend on our purpose in defining it. For the study of physical laws within the transducer, it is usually necessary to study very small systems. For the study of overall performance, it is useful to study the whole system. The relatively new study of systems analysis or systems engineering is devoted to analyzing combinations of smaller units and their interactions and combined performance. Most modern measurement techniques which are either very complicated or very expensive are now being designed using a systems approach, at least in part.

3.3 Transducer Performance

A perfect transducer would behave as in Figure 3.1. Real transducers cannot ever duplicate this, of course. No transducer will give an infinitely large signal since infinite energy would be required. A signal can have many properties, and while both the ordinate and abscissa in Figure 3.1 may refer to, say, amplitude, the frequency may be transferred rather poorly. The straight line in Figure 3.2 is also an idealization. Many transducers are nonlinear, as in Figure 3.3. This is occasionally done on purpose. If a transducer is responding to a^2 and we wish to

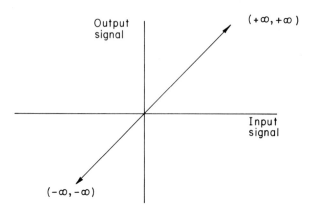

Fig. 3.1. Performance curve for a perfect transducer.

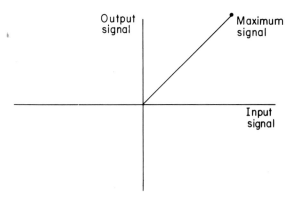

Fig. 3.2. Performance curve for a realistic transducer.

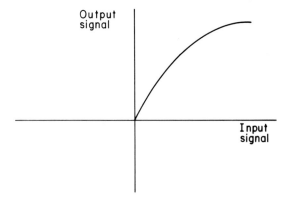

Fig. 3.3. Nonlinear transducer performance curve.

obtain a eventually, it is often possible to arrange a device to measure a^2 and take the square root of the resulting signal internally.

The transducer may have hysteresis, as in Figure 3.4. With this type of response the way in which the value of a signal is approached affects the reading

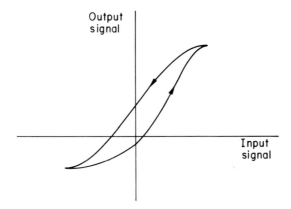

Fig. 3.4. Transducer performance curve with hysteresis (a hysterical transducer).

(or rephrased, the history of the signal is important). There may be *noise* in the transducer. Figure 3.5 illustrates one way in which this could affect the signal.

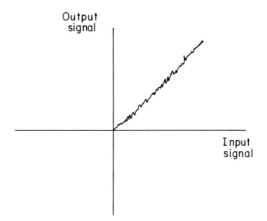

Fig. 3.5. Transducer performance with noise.

The suppression of noise or signal dependent on unwanted variables represents an important part of any experimental design. In radio telescopy, for example, it is possible to obtain information from a signal where the noise is 10 to 100 times larger than the required signal.

One of the most important aspects of transducers is their time dependence. Some instruments function better in the steady state or near steady state, such as a mercury thermometer. Their ability to respond to fluctuations over short periods of time is very poor. This can be an advantage. If we wish to know the average temperature in an air-conditioning duct, the very small fast fluctuations in temperature will be damped out by the thermometer. It can also be a disadvantage. The *manner* in which the thermometer averages may not be appropriate for the purpose in mind, and a false reading may result. Instruments which respond well over short periods of time also have their problems. Oscilloscopes, for example, can have very poor ability to record null or zero readings. Instruments can have harmonic frequencies either on purpose (a tuning fork) or in spite of their designer (a spring balance).

The analytic study of the response of transducers (as well as other systems), and particularly their response with respect to time, is the subject of control theory. This subject is important enough in the subject of measurement that the study of control theory as applied to measurements is sometimes called *general measurement theory*. The first two references at the end of this chapter discuss this topic.

3.4 Monitoring and Control

A type of application of measurement which stands in many ways by itself is the use of a measurement to directly monitor and control the process or events which are being measured. In this case the measurement may or may not be permanently recorded or read. The result is, however, turned back on the process. A simple example of this process is the flyball governor (Figure 3.6). As the drive-

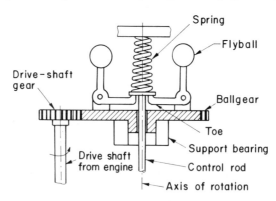

Fig. 3.6. Flyball governor.

shaft gear records by increased revolutions per minute an increase in engine speed, the ball gear is rotated faster and the flyballs are forced outward. This compresses the spring and lifts the control rod which is used to cut off the fuel flow to the engine, slowing it down. The governor is, then, a constant speed device. Notice that an increased speed produces an immediate decrease in speed. The *feedback* to the engine is negative since the signal to the engine is the opposite of the input signal. The analysis of *negative feedback* systems represents a large study within control theory. The problems of even such a simple system may not be obvious. Assume for a moment that the engine is surging (i.e., that the revolutions per minute are fluctuating). If this happens fast enough, the spring may not be strong enough to keep the toe against the top of the control rod. The system can then (if the correct series of events occurs at the correct time) become a *positive feedback* system. This is fundamentally unstable, and the machine will destroy itself. In the author's presence, this unlikely series of events nearly happened during the starting up of a 300,000 horsepower (hp) hydraulic turbine. The watchfulness of the human operator and his quick action in disconnecting the governor and taking over manual control saved the situation. Good examples of positive and negative feedback are seen on a trampoline. High jumps are caused by positive feedback and quick stops by negative feedback.

3.5 Economics and Time

The amount of effort spent on obtaining an answer to a problem includes a consideration of both the amount of time and money available to do the job. The different approaches which this can lead to are best illustrated in the following story.*

> Statement of the problem was simple enough: How long should a 3-pound roast stay in a 325° oven for the center to reach a temperature of 150°F.
> Under pressure, four patterns of attack emerged from four different men in the group of budding engineers.
> There was the Big Project man. He didn't come up with a quick answer, but he had a well-thought-out plan. His idea was to purchase an electric oven with precise temperature control and some first class indicating and recording instrumentation. A series of experiments designed to evaluate the significance of each parameter was outlined. Cost of the project was estimated and a proposal for obtaining a grant from one of the scientific foundations was drafted. With prompt appropriation and reasonable delivery he could have had the answer in six to nine months.
> The Practical Approach appealed to another man. He went out and bought a roast at the local market, an oven thermometer and a meat thermometer at the "five-and-dime." He cooked the roast in a kitchen oven (service environment) and timed it with his wrist watch. While munching medium-rare roast beef sandwiches, he wrote his report.

*C. Carmichael, "Four Kinds of Engineer," *Machine Design*, May 11, 1961. Quoted with permission.

The Supported Judgment man got a quick answer. He reasoned that animal tissue is mostly water and therefore should have about the same specific heat and conductivity. Handbook values gave him the data he needed to apply his basic heat transfer theory. His answer was within minutes of that reported by the Practical Approach man.

The quickest answer came from a student who simply consulted his mother by phone. [the *reference* approach]

Considerations of economics and time require that experiments and measurements often be run at different levels of sophistication. Consider the testing of aircraft wings. A university research worker will test one size of wing to illustrate an improvement he has proposed. For purposes of proving his point, a series of, say, 5 or 10 tests may illustrate the case he wishes to make. If, however, an aircraft company wishes to incorporate this improvement into an aircraft, they may repeat the tests many hundreds of times both in the laboratory and on a test aircraft. The manufacturer must demonstrate that the improvement is indeed an improvement for all cases where the aircraft might work. The manufacturer and researcher are both attempting to demonstrate that the new wing is an improvement. The researcher, however, cannot afford the time for a long series of tests. Furthermore, there is no need to do these tests until a final use for the wing is contemplated. The manufacturer, of course, cannot afford not to make these tests.

3.6 Calibration

Instruments used for measurement may be divided into three classes:

1. *Standards.* The basic measurements of length, mass, time, and voltage by which all other instruments are judged are kept in national standards laboratories. In the United States this is done by the National Bureau of Standards. Other examples are the National Physical Laboratories, Britain; The National Research Council, Canada; and the Australian Research Council. The instruments for comparison of the standards are part of the standard. For example, the balance used in comparing the mass standard with another mass sent for calibration must be the same for every calibration. These standards are considered further in Chapter 4.

2. *Absolute measuring devices.* Some instruments do not require calibration (although this does not imply that they are measuring the required quantity with perfect accuracy). An example of this is the U-tube or manometer.

Figure 3.7 shows a standard manometer whose purpose is to measure a pressure difference. In this case the difference is between an unknown pressure and atmospheric pressure. The pressure will support a water column of height h. The pressure on the unknown side is higher than on the atmospheric side, since the water has been depressed there. The relationship between pressure and height

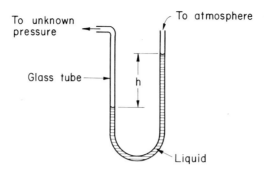

Fig. 3.7. Manometer or U-tube.

is given by

$$P = \gamma h$$

where P = pressure difference
  γ = specific weight of water
  h = height of the water column

Variation in tube diameter does not influence the result. Fluid statics (the study of fluids which are not moving) is well understood, and this equation is extremely accurate if the temperatures of the air and water are constant and the pressure difference is not large. It is unnecessary to calibrate this manometer in the vast majority of applications if γ has been accurately specified. In fact, a manometer is an excellent device for calibrating other pressure-measuring devices. A mercury barometer is a manometer for measuring pressures with respect to vacuum.

Right-angle squares produced by 3, 4, 5 triangles do not require calibration.

It should not be forgotten that the rules and scales used to measure the height of the liquid column in the manometer and the sides of the 3, 4, 5 triangle *are* in need of calibration.

3. *Instruments requiring calibration.* Instruments requiring calibration include the vast majority of practical engineering instruments. Most electrical and electronic instruments require calibration, as do most chemical instruments and many optical instruments. Calibration can be done by the instrument manufacturer and before, during, and after use of the instrument. In many cases all are required.

All instruments should have *traceability*, which is the ability to trace the calibration back to an absolute standard. For example, a home ruler is manufactured from a press which has been calibrated by micrometers and gage blocks by the manufacturer. These micrometer and gage blocks in turn have been calibrated against high precision standards which will have been compared with reference standards in national laboratories. At each stage a price is paid in decreasing accuracy. The higher the accuracy required, the shorter should be the path to absolute or national standards. A student test of cheap 12-in. plastic and wood rulers against a series of gage blocks produced maximum variations among these rulers of up to 1/8 in.!

The calibration should attempt to duplicate, as closely as possible, the conditions of use of the instrument. A flowmeter is a device for producing an obstruction in a fluid flow and measuring the pressure difference across the obstruction. This pressure difference is a function of the total flow and may be used to measure the total flow. The calibration of this instrument will vary with temperature and pressure. If the fluid is a gas, the calibration will vary if the gas has entrained dust or if the pipe has dirt on the inside surface. The operating condition in this case must be reproduced as exactly as possible.

Calibrations may be static or dynamic depending on whether the instrument is to be used to give static or dynamic measurements. A static calibration is produced by using the standard and the instrument to measure a constant value of the required quantity. Dynamic calibration requires testing the instrument under conditions which vary in time.

In some cases calibration is impossible or at least extremely difficult. An obvious case is in clinical psychology where subjective responses are required. If a subject is asked, for example, to state whether or not he saw a flash of light out of the corner of his eye, it is impossible to "calibrate" the percentage of time when he is telling the truth.

3.7 Exercises

3.1 Attempt to fill in some of the "blanks" in Table 3.1.

3.2. Identify the major input, minor input, and output of:
    (a) a tape recorder,
    (b) an accelerometer,
    (c) a flight level indicator,
    (d) a flyball governor,
    (e) a digital voltmeter,
    (f) an automatic choke,
    (g) a Geiger counter.

3.3. Identify as an active or passive transducer the following instruments:
    (a) telescope,
    (b) weather vane,
    (c) electron microscope,
    (d) tachometer,
    (e) automotive timer light,
    (f) automotive temperature indicator.

3.4. (a) Discuss the viewing of a star through a telescope from the point of view of the statements made in the first paragraph of Section 3.1.

(b) Discuss hysteresis and its causes in a mercury thermometer. How is hysteresis related to frequency response in this case?

3.5. Discuss a home heating thermostat as an element in a negative feedback system.

3.6. Discuss bicycle riding as an example of negative feedback. Explain how it is possible to ride "no hands."

3.7. Briefly describe "minimum" and "maximum" types of tests for the following situations. In each case outline a "quick and dirty" type of experiment suitable for a situation where a fast answer is more important than a deeply considered, broad answer and vice versa.

(a) Optimum forward positioning of the wing on a child's glider.

(b) Determination of the density of sand.

(c) Determination of heat loss from a house.

(d) Determination of optimum chair heights.

3.8. Trace the probable path of calibration back to national standards of the following measurements:

(a) School protractors.

(b) The dimensions of lumber.

(c) Microscope graticules. (Microscope graticules are glass plates placed in a microscope with dimensional markings on them. When the microscope is viewed the markings of the graticule are seen imposed on the object being viewed. This is a common way of sizing small objects.)

(d) The diameters of cylinders in automobile engines.

(e) Distances as measured by surveyors using tapes and triangulation.

(f) Thread sizes of bolts.

3.8 Suggestions for Further Reading

Beckwith, J. G., and Buck, W. L., *Mechanical Measurements,* Addison-Wesley Publishing Company, Inc., Reading, Mass., 1963.

Doebelin, Ernest O., *Measurement Systems,* McGraw-Hill Book Company, New York, 1966.

Eckman, D. P., *Industrial Instrumentation,* John Wiley & Sons, Inc., New York, 1950.

Stein, P. K., *Measurement Engineering,* Stein Engineering Services Inc., Phoenix, Ariz., 1964.

4

Statistics

This chapter outlines the mathematical techniques available for treating measurements. This material enables logical decisions to be made concerning the precision of results or groups of results.

4.1 The Usefulness of Statistics

Statistics is a body of mathematical knowledge dealing with quantities which vary in a random manner. The quantities can be integers, decimal numbers, or complicated functions. We shall not be going into the mathematical studies at all deeply. In addition, we shall consider only one application of statistics: to the treatment of errors.

It is often difficult for students who have no familiarity with experiments or measurements to see the usefulness of statistical methods. In many cases, the argument tends to be along the lines "If you want to measure more accurately, buy a better ruler." While this is perfectly true, there are many cases where this approach is simply not possible. In reading this chapter it may be helpful if the student keeps in mind any of a number of examples of realistic measurement problems, the purpose of this being to remind yourself that more accurate rulers are not always possible. A list of problems which might require statistics of the sort we are going to describe are given below:

1. The measurement of average daily production of a "wonder widget" factory.
2. The measurement of the speed of light to maximum precision.
3. The measurement of the diameter of trees in order to determine the amount of lumber available per acre.
4. The measurement of the height of waves on the Atlantic Ocean.
5. The testing of electrical resistors to see if they meet the manufacturer's specifications.

6. The measurement of the average flow rate in a river.
7. The measurement of the number of cars which use a particular road intersection at various times of day.

In each of these examples, a little thought will show you that improvements in measuring instruments will not answer a number of questions for which you would require answers.

In earlier sections we discussed how errors arise and found that the results of measurements are inherently variable. We are, therefore, faced with the problem of attempting to extract the maximum amount of information from assemblies of data which do not give exact answers. There are two properties of assemblies or groups of data which will concern us: first, a tendency of most data to group themselves around a central value; second, the tendency of most data to disperse themselves about this value. We shall be almost exclusively concerned with putting accurate values to these two properties and illustrating their usefulness.

As with any new subject we begin by defining the vocabulary which we shall need. A basic vocabulary is given below. Note that the words may be familiar but that for statistical purposes they are rather more narrowly defined than in a dictionary.

1. *Magnitude.* The objective of most measurements is the magnitude or *size* of a variable.
2. *Observation.* This is a single outcome or result from applying the measurement device—a reading from the device.
3. *Parent population.* If the measurement were made an infinite number of times, we would have a collection of all possible measurements called a parent population. This is obviously a hypothetical or theoretical concept.
4. *Sample.* A sample is a finite group of measurements (or a single one) which comes from the population of all possible measurements. Note that the number of possible samples is also infinite.
5. *Error.* This has been previously defined as the difference between the true value and the measured value. Except in artificial cases the true value is never known. Therefore, the value of an error is never known. If it were, we would correct our data and no longer have an error arising from that source. In artificial cases it is possible to know the value of errors. If, for example, we attempt to calculate the value of π by measuring first the circumference and then the diameter of a circular disk, we can find the errors in the measurement to any degree of accuracy we please.
6. *Accuracy.* This is the closeness to the true value and, like an error, is never known in practical cases.
7. *Precision.* This is the closeness of grouping of data.

The difference between accuracy and precision can be made clear by the following analogy. If, instead of measuring, we shoot arrows at a target, we might get the three possible results shown in Figure 4.1. The diagram at the top represents a set of shots which were precise but inaccurate and that at the bottom represents shots which were accurate but imprecise. In the measurement situa-

Target no. I.
Precise but not accurate

Target no. 2.
Accurate and precise

Target no. 3.
Accurate but not precise

Fig. 4.1. Concepts of accuracy and precision.

tion, the targets would be removed. Then we could say something about precision but not accuracy. In fact, statistics will enable us to do just this and no more.

Errors may be biased or unbiased and give results which are inaccurate or imprecise, respectively. Simple examples are easy to give. A biased error would result from a mistake in calibrating a thermocouple or from a piece of dirt under a micrometer head or from a crack in a thermometer. Unbiased errors may arise in two ways. A single source of error may give them. For example, the torque on the spindle of a micrometer may be varied from one measurement to the next. The error of interpolating a reading on a dial with large divisions can be alternately positive or negative for particular readings. Second, unbiased errors can also arise as a combination of many sources. A micrometer reading can be in error due to dirt, spindle torque, temperature, and operator changes at the same time.

For the purpose of statistical analysis you should note that we can give the name *error* to a measurement which superficially is a true value. If we wish to measure the average production of car engines per day from a factory, then the difference between a single day's production and the average would be called

an error in the statistical sense. The ambiguity is a result of our artificial definition of the average production as a "true" average.

4.2 Presentation of Results

Suppose that we begin with a set of repeated measurements of some
quantity. For the present let us assume that they have been repeated because we
know they vary and we want an average. The simplest way to present the data
would be to list them in order of increasing magnitude. An example is given in
Figure 4.2. If we further grouped this array into, let us say, seven groups, then
we would have a frequency distribution like the one in Table 4.1. The added

32	51	64	76
37	53	64	77
41	55	68	78
43	57	69	82
43	57	69	88
48	60	71	88
50	61	75	91
51	63	75	94

Total number of students = 32

Fig. 4.2. Table of student marks.

advantage of the frequency distribution over the array of data is that the central
tendency is more clearly brought out. In the array we might assume that the data
are grouped about some value halfway between the highest and lowest value but
we have no way of knowing. In the frequency distribution we can see visually
that there are, in fact, more values near the middle of the range. This fact can be
brought out even more clearly in a histogram, which presents the same information in a completely graphical form. The example data are given in a histogram
in Figure 4.3.

Table 4.1.   Frequency distribution.

Class interval	Observations in class	Total
30 – 39	//	2
40 – 49	////	4
50 – 59	///////	7
60 – 69	////////	8
70 – 79	//////	6
80 – 89	///	3
90 – 100	//	2

Check = 32

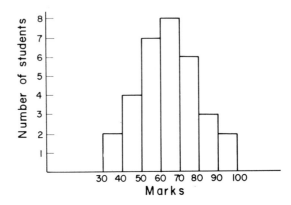

Fig. 4.3. Histogram of student marks.

We can now see very clearly the tendency of the data to group themselves in a central region or to display a central tendency. The data dispersed about this central region become more infrequent as we go further from the middle range.

The information given in Figures 4.2 and 4.3 and Table 4.1 can be plotted graphically in many other forms, all of them giving essentially the same information. Figure 4.4 is a frequency polygon. Figure 4.5 is the same data plotted, but

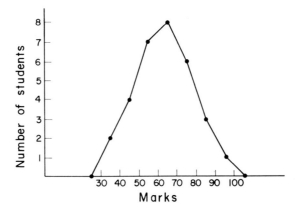

Fig. 4.4. Frequency polygon.

in this case the number of classes have been doubled and the range has been cut in half. The range is the range of values put in a single class or group. In Figure 4.3 the range is 10, and in Figure 4.5 the range is 5. The data can be plotted as a *relative frequency diagram* (Figure 4.6). In this type of plotting, the frequency (the abscissa of the histogram) has been divided by the total number of measurements taken. The vertical scale is now relative occurrence or percentage of the

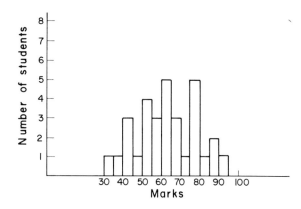

Fig. 4.5. Histogram-class width is now 5.

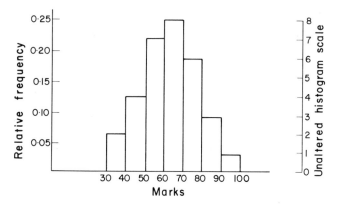

Fig. 4.6. Relative frequency diagram.

total. The relative frequency diagram is, then, just a histogram with the numbers on the vertical axis changed.

We could present the same diagram as a *cumulative frequency diagram*. This is best described as a "less than" diagram. The quantity plotted is the total number of readings (or proportion of readings) which have a value less than the value on the ordinate of the graph. The data of the previous figures have been arranged in this form in Figure 4.7.

4.3 Analysis of Central Tendency

There are several ways of describing a typical or central value of a group of data. We shall discuss the median, mode, geometric mean, and arithmetic mean. The purpose of each is to produce a typical or representative value for all the

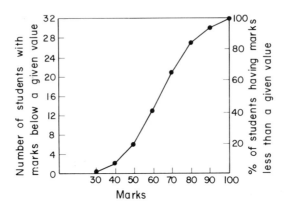

Fig. 4.7. Cumulative frequency diagram.

data. The particular type used depends on the purpose to which the value is to be put.

Median. The median is the middle value. Half the data will lie above this value and half below. It is not the value halfway between the largest and smallest value. For example, with the numbers 1, 2, 3, 4, and 5 the median is 3. With the numbers 2, 5, 10, 10, and 15 the median is 10. The median is less affected by extreme values than other types of central measures. For example, let us assume that we have five engineers with average yearly salaries of $10,000, $11,000, $12,000, $12,500, and $25,000 per year. The average salary, obtained by adding all salaries and dividing by 5, is $14,000. This is not a good measure of a typical value because no engineer makes within $1600 of this figure. The problem is that the single extremely high value of $25,000 has influenced the mean too much. The median, though, is $12,000, which is a quite typical value and a realistic estimate of what four out of the five engineers make.

The concept of a median can be generalized to the idea of quantiles. An example of a quantile is a quartile. An upper quartile is a value above which one quarter of the values of the data lies. Conversely, a lower quartile is the value below which one quarter of the data lies. The use of quartiles then implies that the data are divided into ranges, each of which contains one quarter of the data. A decile is a range containing one tenth of the data. It is usual to consider only upper and lower quartiles and deciles. An example of the use of deciles and quartiles is given in Figure 4.8.

Mode. The mode is the most frequent value of the data. This would be the peak of a histogram and the value would be taken as the mid-value of the class or range which is most frequent. It is possible to obtain histograms with more than one peak. There are called bimodal distributions. Bimodal distributions arise if, for example, samples come from two different populations.

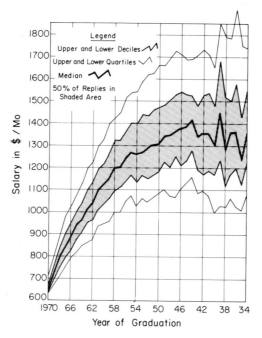

Fig. 4.8. Salaries of professional engineers in Canada in July 1970 (From *Engineering Digest*, Oct. 1970; reproduced with permission).

Harmonic mean. The harmonic mean is calculated according to the following formula:

$$\text{H.M.} = \frac{n}{\Sigma\,(1/x_i)} \tag{4.1}$$

H.M. is the harmonic mean, n is the total number of data points considered, and x_i is the set of particular values of the data. Σ is the summation convention as used in calculus and implies summation over all values of n. The harmonic mean of the values of 1, 2, 3, and 4 is then

$$\text{H.M.} = \frac{4}{1/1 + 1/2 + 1/3 + 1/4} = \frac{4}{2.08} = 1.92$$

The usefulness of the harmonic mean is in obtaining typical values of rates and speeds. If the numbers in the above example were taken to be speed in miles per hour, then if an object went 1 mph (mile per hour) for 1 mile, then 2 mph for 1 mile, etc., then the average speed over the total distance would be 1.92 mph. It is left to the student to check that this is true. You should note that the distances traveled and not the times are constant for each interval.

Geometric mean. The geometric mean is used for things which grow as a geometric progression or might be expected to grow proportionally to their size. The geometric mean is calculated by means of the formula

$$\text{G.M.} = \sqrt[n]{x_1 \cdot x_2 \cdot x_3 \cdot x_4 \cdots x_n} \qquad (4.2)$$

For example, let us suppose we have a share in a company which has the values given in Table 4.2 over 3 yrs.

Table 4.2

Year	Value	Value as a Proportion of Previous Year
1	100	
2	50	0.50
3	100	2.00

The last column gives the multiplying factor by which we multiply one year's value to get the next year's value. It therefore represents the rate of growth of the share. The average growth would be incorrectly given as $\frac{0.50 + 1.00}{2}$ (the arithmetic mean) since this would give 1.25 and this would be an incorrect value for year 3 (i.e., $100 \times 1.25 \times 1.25 \neq 100$). The correct value is the geometric mean. This is given by

$$\text{G.M.} = \sqrt[2]{0.50 \times 2.00} = 1.0$$

and $100 \times 1.0 \times 1.0 = 100$, which is the correct value for year 3. A less trivial example is the following: The population of a small city is given in Table 4.3.

Table 4.3

Year	Population	Increase	Growth Rate
1966	29,894		
1967	31,422	1528	1.05111
1968	32,527	1107	1.03517
1969	33,349	822	1.02527
1970	34,681	1332	1.03999

$$\text{Average growth rate} = \sqrt[4]{1.0511 \times 1.0352 \times 1.0253 \times 1.0399}$$
$$= \sqrt[4]{1.16013} = 1.03783$$

Therefore, average growth is 3.78%.

Arithmetic mean. The arithmetic mean is the most common and most useful of the averages. It is simply the sum of all values divided by the number of values. The formula then is

$$\text{A.M.} = \bar{x} = \frac{\Sigma\, x_i}{n} \tag{4.3}$$

We shall use the symbol $\bar{x}$ to represent this mean. It is possible to simplify the calculation of $\bar{x}$ in certain circumstances. If we must average 7.01, 7.02, 7.03, 7.04, and 7.05, we may subtract the 7 from all values before beginning the calculation and add it on at the end. Therefore,

$$\frac{0.01 + 0.02 + 0.03 + 0.04 + 0.05}{5} = \frac{0.15}{5} = 0.03$$

$$\bar{x} = 7.00 + 0.03 = 7.03$$

This method is, of course, most useful when large samples are being considered. Another method of calculation of $\bar{x}$ is given in Section 6.11.

4.4 Analysis of Dispersion

The two histograms in Figure 4.9 have the same mean. The precision (closeness of grouping) of the samples, though, is quite different. We now require a numerical measure of this dispersion. As with means, there are several ways of presenting this information. The simplest would be to present the maximum and minimum values obtained, that is, the range of data. It is easy to see why this method is seldom satisfactory. The weight of decision as to the amount of spread is placed on the most extreme values and there is every reason to believe that these values should carry very little value in our minds since there is the most likelihood of these values being "wrong." Also, the values of the range will probably increase with the size of the sample.

Mean deviation. Occasionally a measure of dispersion is calculated according to the formula

$$d = \frac{\Sigma\, |x_i - \bar{x}|}{n} \tag{4.4}$$

That is, the deviations of each data point from the arithmetic mean are evaluated, the absolute value is taken, and the average of these is calculated. The absolute value must be taken since the quantity

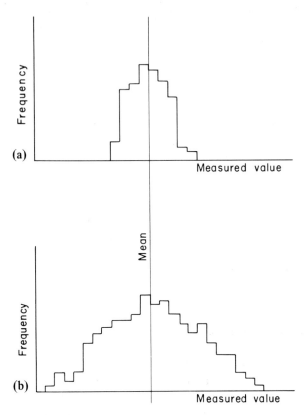

Fig. 4.9(a) Histogram with small dispersion.
       (b) Histogram with large dispersion.

$$\frac{\Sigma\,(x_i - \bar{x})}{n} = \frac{\Sigma\,x_i - \Sigma\,\bar{x}}{n} = \frac{\Sigma\,x_i}{n} - \frac{n\bar{x}}{n} = \bar{x} - \bar{x} = 0$$

This allows us to give another meaning to the arithmetic mean as the value where the sum of the deviations of the data from it is a minimum and in fact equal to zero. The mean, then, is the center of gravity axis of the histogram. That is, the histogram is balanced about a vertical line drawn through $x = \bar{x}$ if each data point has equal "weight."

The mean deviation given by equation (4.4) is very seldom used in statistical calculations.

Variance. The variance of a sample is a measure of the dispersion, and its form may be obtained by the following argument. The deviation of a single data point from the mean is also called a residual and is equal to $x_i - \bar{x}$. These values may be positive or negative and, as we have just proved, their average is zero. If we square these residuals, the results will all be positive, and we may take the average of the squares. The result is called the variance (s^{*2}) and is then calculated as

$$s^{*2} = \frac{\Sigma (x_i - \bar{x})^2}{n} \tag{4.5}$$

This is sometimes called the mean square deviation since the mean of the square of the deviations is taken. This quantity is most important in statistics, and we shall be using it many times.

Standard deviation. The standard deviation of a sample is simply the square root of the variance or s^*.

$$s^* = \sqrt{\frac{\Sigma (x_i - \bar{x})^2}{n}} \tag{4.6}$$

This, for obvious reasons, is often called the root mean square deviation, abbreviated the rms* deviation. For students with a knowledge of dynamics the standard deviation is entirely analogous to the radius of gyration of a histogram about the mean.

One of the advantages of the standard deviation over the variance is that the units of the standard deviation are the same as the units of the quantity measured. That is, the standard deviation of a measurement in inches will also be expressed as inches, whereas that of the variance will be expressed as inches squared.

Each of the methods of characterizing dispersion results in a statistic. A *statistic* is a value which is descriptive of, or characteristic of, the data. We shall make use later of the idea that means, dispersions, and any other quantities calculated from the data are statistics.

The calculation of variance and standard deviation can be quite laborious if done by hand. Modern desk calculators and computers make the job very easy. If it is necessary to compute s^* by hand, a table should be used, as in the example at the end of this section.

In addition, formula (4.6) may be rewritten as

$$s^* = \frac{1}{n} \sqrt{n \Sigma x_i^2 - (\Sigma x_i)^2} \tag{4.7}$$

*This terminology is more frequently used in sampling of continuous or analog information.

To prove the identity of equations (4.6) and (4.7) consider the quantity under the square root in equation (4.6):

$$\frac{\Sigma (x_i - \bar{x})^2}{n} = \frac{\Sigma (x_i^2 - 2x_i\bar{x} + \bar{x}^2)}{n}$$

$$= \frac{\Sigma x_i^2 - 2\bar{x} \Sigma x_i + \Sigma \bar{x}^2}{n}$$

$$= \frac{\Sigma x_i^2}{n} - \frac{2\bar{x} \Sigma x_i}{n} + \frac{n\bar{x}^2}{n}$$

since $\Sigma \bar{x}^2 = n\bar{x}^2$. Further, this is equal to

$$\frac{\Sigma x_i^2}{n} - 2\bar{x}^2 + \bar{x}^2$$

Therefore,

$$\frac{\Sigma (x_i - \bar{x})^2}{n} = \frac{\Sigma x_i^2}{n} - \bar{x}^2$$

Placing this in equation (4.6), we obtain

$$s^* = \sqrt{\frac{\Sigma x_i^2}{n} - \bar{x}^2}$$

$$= \frac{1}{n} \sqrt{n \Sigma x_i^2 - n^2 \bar{x}^2}$$

$$= \frac{1}{n} \sqrt{n \Sigma x_i^2 - (\Sigma x_i)^2}$$

since $\bar{x}^2 = \left(\frac{\Sigma x_i}{n}\right)^2$. Q.E.D.

Here is an example illustrating the use of equations (4.6) and (4.7).

Example

Given a set of data we shall calculate s^* using both methods. The data are

3.416, 3.412, 3.413, 3.412, 3.415, 3.413, 3.411, 3.414, 3.414, 3.413

According to equation (4.6), we may proceed as shown in Table 4.4.

Table 4.4

1	2	3	4
x	$(x - 3.41) \times 10^3$	$x - \bar{x}$	$(x - \bar{x})^2$
3.416	6	+0.0027	0.729×10^{-5}
3.412	2	−0.0013	0.169×10^{-5}
3.413	3	−0.0003	0.009×10^{-5}
3.412	2	+0.0013	0.169×10^{-5}
3.415	5	+0.0017	0.289×10^{-5}
3.413	3	−0.0003	0.009×10^{-5}
3.411	1	−0.0023	0.529×10^{-5}
3.414	4	+0.0007	0.039×10^{-5}
3.414	4	+0.0007	0.049×10^{-5}
3.413	3	−0.0003	0.009×10^{-5}
	$\Sigma = 33$		$\Sigma = 2.01 \times 10^{-5}$

Column 2 is used to make the calculation of $\bar{x}$ easier, as outlined in Section 4.3.

Therefore, the total of column 2 = 0.033 and the average of column 2 is $\frac{0.033}{10} = 0.0033$. Therefore,

$$\bar{x} = 3.41 + 0.0033 = 3.4133$$

We now fill in columns 3 and 4. From the bottom of column 4 we may calculate

$$s^* = \sqrt{\frac{2.01 \times 10^{-5}}{10}} = 0.001417$$

If we proceed according to equation (4.2),

x	x^2
3.416	11.6691
3.412	11.6417
3.413	11.6986
3.412	11.6417
3.415	11.6622
3.413	11.6486
3.411	11.6349
3.414	11.6554
3.414	11.6554
3.413	11.6486
$\Sigma x = 34.133$	$\Sigma x^2 = 116.506$

$$s^* = 1/10 \sqrt{10 \times 116.504 - (34.133)^2} = 0.001414$$

Either form of the calculation may be used depending on the circumstances and the convenience of application. It should also be noted that the difference of two nearly equal quantities is taken in this last calculation. In these circumstances, large errors can arise if the calculations are rounded too soon. This problem is discussed in Section 4.8. This accounts for the discrepancy in the two answers in the sixth decimal place. A graphical method of obtaining $s*$ is given in Section 6.11.

Coefficient of variation. This measure is simply a method of expressing the standard deviation (which is in the same units as the arithmetic mean) as a percentage of the mean. Hence,

$$C.V. = \frac{s*}{\bar{x}} \times 100\% \qquad (4.8)$$

The coefficient of variation can sometimes be misleading if the zero value of the measurements is arbitrary. This is true in temperature measurements so that a co-efficient of variation based upon identical data will differ depending on whether it is based upon the Fahrenheit or Celsius scales. This will be true even though the C.V. is not in either Fahrenheit or Celsius units.

4.5 Theoretical Distribution Curves and Populations

The basic vocabulary of statistics included a definition of a parent popula-tion as the total of all possible measurements. If we were to take larger samples and plot them as relative frequency diagrams, we would find it natural to increase the number of classes on the diagram as the sample gets larger. We would then have a series of situations as shown in Figure 4.10. As the number of classes goes to infinity the class width will go to zero. In the limit the relative frequency will result in a smooth curve (but not the histogram since each class would still con-tain an infinite number of points and therefore be off the top of any histogram). The resulting smooth curve is a *theoretical distribution curve.* It has the great advantage that it may be treated analytically, unlike the relative frequency dia-gram. We can, for example, differentiate and integrate this curve. This curve also represents the population of all possible measurements (but not the "true" value). We can see then that *we know as much as we can possibly know about the meas-urement if we know the properties of this curve. The finite sample we are always forced to take can be considered as an attempt to find the properties of this curve.*

It would be a very difficult world indeed if the theoretical distribution curve for every situation were different. In fact, a very few types of curves have been shown to explain a wide range of phenomena. A few examples are given

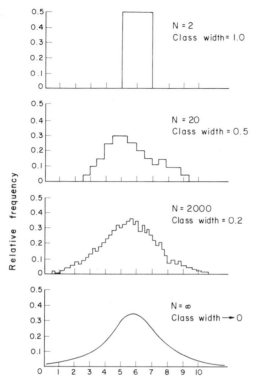

Fig. 4.10. Effect of increasing sample size and decreasing class width.

below. We cannot *prove* that our particular problem is described by a particular curve except in uncommon situations.

1. Binomial distribution. This distribution is expected when we have an either-or situation, for example, heads or tails, black balls or white balls, etc., and the probability of obtaining either is equal. (Probability is discussed in Section 4.6.) If we threw four coins at once, the binomial distribution would predict the probability of obtaining

1. Four heads and no tails.
2. Three heads and one tail.
3. Two heads and two tails.
4. One head and three tails.
5. No heads and four tails.

2. Poisson distribution. This distribution is used in discussing isolated events affected by chance alone and not influencing each other. The nonoccurrence of the event has no meaning in this case. Examples are cars passing a given point, clicks on a Geiger counter, and the number of bacteria in a microscope field of view. The distribution would predict the relative frequency of no cars in a 5-min period, and one car, two cars, three cars, etc., in the same 5-min period.

3. Normal or Gaussian distribution. This is the distribution which has been found to describe errors or variations in members of a single population. This is the only distribution which we shall consider in detail.

The equation for this distribution is given as

$$y = \frac{1}{\sigma\sqrt{2\pi}} \cdot e^{\frac{-(x-X)^2}{2\sigma^2}} \tag{4.9}$$

where y = relative frequency of x, the measured variable
      X = mean value of x, for the population
      σ = standard deviation of x, for the population
      σ^2 = variance of x
      e = 2.71828. . .

The form of this theoretical distribution curve is shown in Figure 4.11. It should be noted that this theoretical distribution curve is completely defined by

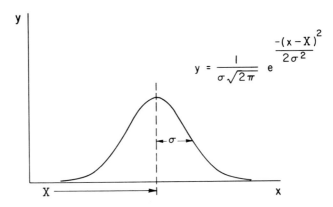

Fig. 4.11. Gaussian or normal distribution: X = mean value, σ = standard deviation.

two parameters, the mean X and the standard deviation σ. That is, it may be plotted if X and σ are known. This theoretical distribution then has similar properties to our samples, i.e., a mean and a standard deviation. Unlike the theoretical curve, however, the sample is not completely determined by these two quantities.

It should be noted that when this theoretical distribution curve was developed in 1773 it was known as the *law of errors* because it was found to represent the errors of observation in astronomy and the other physical sciences remarkably well. One could, therefore, expect that the observations generated by a measurement device would be normally distributed, providing the observations were the resultant of many perturbing factors each of which exerted a small influence. The broad applicability of the normal distribution is further extended by means of the central limit theorem. The central limit theorem may be stated in the following way:

1. Assume a set of values of something which is *not* normally distributed.
2. Draw a sample from this population and calculate the average of the sample.
3. Obtain a large group of averages by obtaining a large group of samples.
4. The central limit theorem states that the *averages* of the various samples *will* be normally distributed even if the population and the samples themselves are not. A proof of this theorem requires a suitable knowledge of calculus and will be found in any comprehensive book on statistics.*

Since an infinite number of measurements are required to define a curve which is completely defined by two numbers (X and σ), we may conclude that X and σ may never be known exactly. They can, however, be estimated. We want to do this since the curve represents the best possible measurement set (an infinite number). The method of estimating X and σ is explained in Section 4.7.

4.6 Probability

Probability theory is a branch of mathematics that is concerned with conceptual experiments in which the outcomes are random events and which can be performed, at least in principle, a large number of times. We shall use it here to enable us to determine the relationship between samples and populations and to give some further ways of interpreting the concept of a population.

If an event can occur in a ways out of a total of n possible equally likely ways, then its probability of occurrence is given by

$$P[E] = \frac{a}{n} \tag{4.10}$$

The probability of an event is a number between 0 and 1. If an event *cannot* occur, then its probability is 0. If an event *must* occur, then its probability is 1.

*For example, see I. Guttmann and S. S. Wilks, *Introductory Engineering Statistics,* John Wiley & Sons, Inc., New York, 1964, p. 295.

Two descriptions of probability are of interest:

1. A priori probability.
2. Relative frequency probability.

A priori probability. The *a priori* probability of an event may be defined by

$$P[E] = \frac{\text{Number of events favorable to } E, \text{ all of which are equally likely}}{\text{Total number of events possible}} \quad (4.11)$$

A priori probabilities may be calculated for particular experiments in which the mechanics of the experiment which generate the random events are understood. Probability theory is appropriate to conceptual experiments and may be used for a basis for isolating any bias that may exist in real-world experiments.

For example, the probability of occurrence of a head in a conceptual experiment involving the flipping of a coin is ½, which is also the probability of occurrence of a tail. This means that if we flipped a coin a large number of times, then we should expect about ½ heads and ½ tails to occur. If our real-world experiment were biased in some way, then we might observe, say, a larger number of heads.

Some examples of a priori probability follow.

Card Drawing

In a deck of playing cards there are four suits each containing 13 cards for a total of 52 cards. There are four face cards in each suit. Consider the following probabilities:

$$P[\text{Two}] = 4/52 = 1/13$$
$$P[\text{King}] = 4/52 = 1/13$$
$$P[\text{Ace of Hearts}] = 1/52$$
$$P[\text{Face card}] = 16/52 = 4/13$$

Die Rolling

A die (plural dice) possesses six faces numbered 1 to 6:

$$P[1] = 1/6$$
$$P[3] = 1/6$$
$$P[\text{Even number}] = 3/6 = 1/2$$

A pair of dice possesses a total of 12 faces, but there are 36 possible combinations of faces that can turn up when a pair of dice is rolled. Figure 4.12(a)

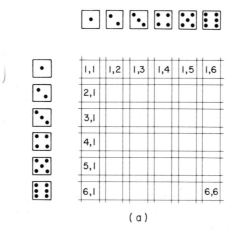

(a)

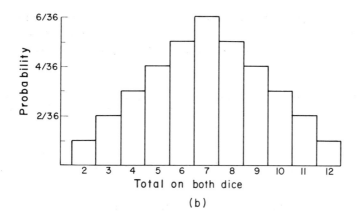

(b)

Fig. 4.12. Probabilities associated with the rolling of a pair of dice.

shows all the combinations that can occur. If we are concerned with calculating the probabilities of occurrence of various sums that may occur on the faces of a pair of dice, then the following probabilities may be calculated:

$$P[2] = 1/36$$
$$P[4] = 3/36 = 1/12$$
$$P[7] = 6/36 = 1/6$$
$$P[12] = 1/36$$

Figure 4.12(b) shows the distribution of the probabilities of occurrence of all possible sums.

Roulette Wheel

A roulette wheel has 37 possible outcomes each time that it is spun, the numbers 1–36 and 0. Some roulette wheels have 00 in addition to 1–36 and 0. Let us consider the first-mentioned wheel.

The probability of any number occurring is 1 in 37 since the total number of possible events is 37. A typical gambling establishment offers odds of 35 to 1 on all numbers except 0. That is, $1 placed on any winning number will yield $35. A patron can, therefore, expect to win

$$\$35 \times 1/37 = 94.6 \text{ cents}$$

over a long-run sequence of bets on the same number. This simple calculation demonstrates why gambling establishments make money over the long run from roulette and other games of chance.

Three examples of a priori probability calculations have been described above. The a priori description of probability is useful for developing understanding of the concept of probability but has only limited applications to measurement.

Relative frequency probability. The probability of an event is equal to the limit of the relative frequency as the total number of observations tends to infinity, i.e.,

$$\lim_{n \to \infty} \frac{f(i)}{n} = P(i) \tag{4.12}$$

This statement is known as the relative frequency postulate.

Consider a container which holds 100 balls of five colors with the distribution of numbers shown in Table 4.5.

Table 4.5

Color	Number	Relative Frequency
White	40	0.40
Blue	20	0.20
Red	20	0.20
Yellow	10	0.10
Green	10	0.10
Total	100	

The relative frequencies shown in Table 4.5 are calculated from dividing the number of balls of each color by the total number of balls. That is, the relative frequency of the white balls is calculated from 40/100.

If the balls were sampled randomly one at a time, with each ball being replaced after a drawing, then the following statements could be made:

Probability of drawing a white ball = 0.40

Probability of drawing a red ball = 0.20

and so on. That is, if a large number of balls were drawn, with each ball being replaced after a drawing, then we could expect about 40% of the drawings to be white balls, 20% blue balls, and so on.

The student mark data described in Section 4.2 may be thought of in an analogous manner. For example, assume that the marks shown in Figure 4.6 represent stable relative frequencies characteristic of the particular group of students from which this class is drawn. If we continued to examine this class again, then the probability of occurrence of a mark being obtained in the interval 40-50% is 4/32 = 0.125; the probability for the 60-70% interval is 0.250; and so on. In other words, the total group may be conceived in terms of a population which obtains marks in the proportions indicated by Figure 4.6.

It is convenient to think of measurement problems in a similar manner. A container can be considered to represent a measurement device along with the unknown magnitude of a variable of interest. The container may be thought of as containing a parent population of observations with a specific mean and standard deviation. The population mean represents the true value of the magnitude of the variable of interest that an observer is attempting to determine. The standard deviation of the observation reflects the precision of the particular measurement device. The problem facing the observer is to estimate the population mean and standard deviation from a limited sample of observations drawn from this container.

4.7 Prediction of Population Properties

With the population distribution as our required information and the interpretation of its meaning available from the previous section, we now wish to use our sample to predict the population mean and standard deviation.

The *standard form* of the normal distribution is given by

$$y = \frac{1}{\sqrt{2\pi}} \cdot e^{-z^2/2} \tag{4.13}$$

where $z = (x - X)/\sigma$.

Equation (4.13) may be obtained from equation (4.9) by substituting the above expression for z and putting $\sigma = 1$. This change will yield a variable with a mean of zero and a variance of unity. This amounts to a shift in origin to the center of the distribution and a change of scale so that one standard deviation becomes the new unit of measure on the axis of values of the variable.

The standardized form of the normal distribution is also normal. Figure 4.13 is a graph of the standard form of the standardized normal curve. The total

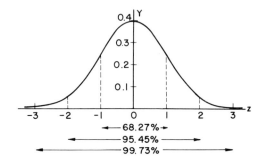

Standard form of the normal distribution

$$y = \frac{1}{\sqrt{2\pi}} \, e^{-z^2/2}$$

where $z = (x - X)/\sigma$

Fig. 4.13. Standard form of the normal distribution.

area under this curve is 1. Note that it can be proved that the following proportions of the area are contained within the bands of z indicated below:

$z = \pm 1$ contains 68.27% of the area under the curve

$z = \pm 2$ contains 95.45% of the area under the curve

$z = \pm 3$ contains 99.73% of the area under the curve

Tables of the area under the standard form of the normal curve for a variety of z values are available. Table I.1 in Appendix I provides such a tabulation of areas. The use of Table I.1 may be illustrated by a simple example. Consider a normal parent population with mean $X = 150$ units and standard deviation $\sigma = 10$ units, and let us calculate the proportion of the observations between 135 units and 175 units.

$$z = \frac{x - X}{\sigma} = \frac{135 - 150}{10} = \frac{-15}{10} = -1.5$$

and

$$\frac{175 - 150}{10} = \frac{25}{10} = 2.5$$

which indicates that $z = -1.5$ and 2.5. Table I.1 indicates that 0.4332 of the area of the curve is within the band $z = -1.5$ and 0 and that 0.4938 of the area is within the band $z = 0$ to 2.5. In other words, 92.70% of the area of the parent distribution defined above falls within the interval of 135 to 175 units.

Let us further assume that the above parent population has been estimated from a large number of observations of the length of a specimen which is 150 mm in length. The standard deviation of 10 mm would be indicative of the precision of the measurement process used. Now, since we have calculated that 92.70% of all observations fall within the band 135-175 mm, the relative frequency postulate allows us to make the statement "The probability of obtaining a new observation with a magnitude between 135 and 175 mm is 0.927."

Let us now calculate the following probabilities with the aid of Table I.1 using $X = 150$ and $\sigma = 10$:

$$P[155 \leq x \leq 160]$$
$$z = (155 - 150)/10 \text{ and } (160 - 150)/10 = 0.5 \text{ and } 1.0$$

From Table I.1, for $z = 0.5$, area = 0.1915, and for $z = 1.0$, area = 0.3413. Therefore,

$$P[155 \leq x \leq 160] = 0.3413 - 0.1915 = 0.1498$$
$$P[125 \leq x \leq 128]$$
$$z = (125 - 150)/10 \text{ and } (128 - 150)/10 = -2.5 \text{ and } -2.2$$

From Table I.1, for $z = -2.5$, area = 0.4938, and for $z = -2.2$, area = 0.4861. Therefore,

$$P[125 \leq x \leq 128] = 0.4938 - 0.4861 = 0.0077$$

These techniques enable us to use the population properties to predict the sample properties. This, however, is the reverse of the usual problem.

The prediction of population properties requires, as has been stated, an estimation of X and σ. It is natural to assume that the best estimates are $\bar{x}$ and s^*, the sample mean and standard deviation. $\bar{x}$ *is the best estimate of* X. *However,* s^* *is not the best estimate of* σ.

The best estimate of σ is given not by s^* but by the quantity

(4.14)

$$s = \sqrt{\frac{\sum\limits_{1}^{n} (x - \bar{x})^2}{n - 1}}$$

This equation should be compared with the equation for s^*. In equation (4.14) the denominator contains the term $n - 1$ and not n. $n - 1$ represents the *number of degrees of freedom* associated with the estimate *s* of the standard deviation σ of the parent population.

The number of degrees of freedom associated with a sample statistic is defined as the number *n* of the independent observations in a sample minus the number *k* of sample statistics already calculated from the observations, i.e.,

$$\nu = n - k$$

The number of degrees of freedom associated with the estimation of *s* is $n - 1$ since the sample statistic $\bar{x}$ has already been estimated from the data.

The concept of degrees of freedom frequently gives considerable difficulty to students new to the subject. The implications can be made clear if we attempt to calculate the standard deviation of a single measurement. For the sample, if the measurement was, say, 3.0, then

$$s* = \sqrt{\frac{(3.0 - 3.0)^2}{1}} = 0$$

i.e., there is no spread for the sample, which is correct. However, for the *population* we use

$$s = \sqrt{\frac{(3.0 - 3.0)^2}{0}} = \frac{0}{0}$$

which is indeterminate. This is also correct since the mathematics is saying that we have no information about the spread of the population.

We may consider degrees of freedom in another way. For a sample of size 10, say, we have 10 separate independent pieces of information. Once we have calculated a mean for this set of 10, then we do not have 11 independent pieces of data since the mean is just a rearrangement of the data. The concept of degrees of freedom is a way of ensuring that 11 pieces of data are not calculated using 10 original pieces of data. For our sample of 1 data point above, it shows us that we cannot calculate a mean and a standard deviation from a single piece of data.

The relationship between *s** and *s* may be written as

$$s = \sqrt{\frac{n}{n-1}} \cdot s* \tag{4.15}$$

The quantity $\sqrt{\frac{n}{n-1}}$ is often called Bessel's correction. A derivation of Bessel's correction is given in Appendix II.

Equation (4.15) illustrates that as *n* gets larger, the difference between *s** and *s* gets smaller. That is, as the sample gets larger, the standard deviation of the sample (*s**) becomes a less biased estimate of the population standard deviation.

We frequently wish to plot on a histogram the best-fitting normal curve. This will be the normal curve with the same mean and standard deviation as the sample. Equation (4.9) should then be used in the form

$$Y = \frac{ni}{\sigma\sqrt{2\pi}} \cdot e^{-(x-X)^2/2\sigma^2} \tag{4.16}$$

where n = total number of observations in the sample
$\quad\;\; i$ = width of the class interval used.

The area under the curve then becomes ni square units. Of course, the histogram could be converted to a relative frequency diagram and the horizontal scale changed to units of $z = \frac{x-X}{\sigma}$. The standard form of the normal distribution [equation (4.13)] would then plot as a best fit on this diagram.

4.8 Sampling Distributions

It has been suggested above that the basic measurement problem is to estimate the parameters of the parent population associated with a particular measurement device when it is used to establish the magnitude of a particular variable.

Figure 4.14 shows a simple measurement device consisting of a dial gage mounted on a stand with a reference plate and which is set up to measure the length of metal specimens. For the specific specimen shown we wish to establish the parameters X and σ of the parent population. X represents the true length of the specimen, and σ represents the precision of the particular device. If we generated an infinite number of observations, then we would not have to concern ourselves with σ since we would know X exactly.

If the particular measurement device shown in Figure 4.14 were replaced by a second device, say a micrometer, then we could expect the magnitude of σ to change. We would not expect X to change for the particular specimen unless a biased error existed in either of the measurement devices.

X cannot be observed directly, and we must infer something about it from readings R displayed by our measurement device. In addition, we generate only a limited number of observations and may calculate the sample statistics $\bar{x}$ and s from these observations. Figure 4.14 indicates that there is a difference between $\bar{x}$ and X. This does not represent a biased error but simply reflects the statistical fluctuation inherent in sample statistics because of the limited data upon which they are based.

A random sample of observations is obtained by a procedure in which each member of a population has its correct probability of being included in the sample of observations.

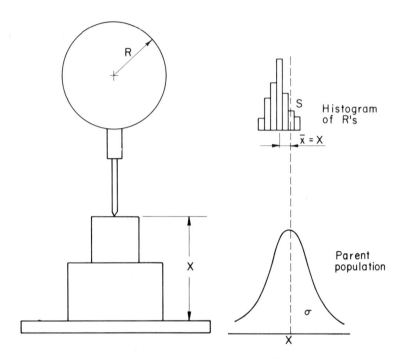

Fig. 4.14. Simple measurement problem.

In other words, the relative frequency of occurrence of observations within a sample simply reflects their probability of occurrence in the parent population and is independent of the characteristics of the process used to generate the observations. The statistical theory of sampling discussed here is based upon the premise that the sample of observations is a random sample.

A sampling distribution describes the frequency distribution of sample statistics such as $\bar{x}$s obtained from samples of n observations drawn from a parent population.

Figure 4.15 indicates that for any practical parent population we may obtain a series of samples of n observations and calculate the means $\bar{x}_1$, $\bar{x}_2$, . . . of each of these samples. These means, or sample statistics, will exhibit some variation around the population mean X, and the extent of this variation may be calculated from statistical theory.

The sample statistics $\bar{x}_1$, $\bar{x}_2$, . . . will tend toward a normal distribution as n increases. For a normal parent population, the distribution of the sample statistics will be normally distributed. This sampling distribution is called the *sampling distribution of the sample means*.

The sampling distribution of the sample means has a mean value equal to the population mean value X and a standard deviation equal to $\sigma/\sqrt{n}$. The standard deviation, $\sigma_{\bar{x}}$, of the sampling distribution is usually called the standard error when dealing with measurements.

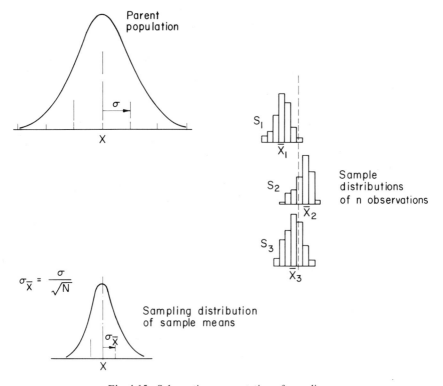

Fig. 4.15. Schematic representation of sampling.

This statement is correct for large or small samples when the parent population is normal. The sampling distribution of sample means is essentially normal for $n \geq 10$ even when the parent population is nonnormal, as the central limit theorem states.

Numerical example. Consider a parent population with a mean $X = 78$ units and standard deviation $\sigma = 12$ units. If we draw samples of $n = 9$ random observations, then

$$\text{Standard error } \sigma_{\bar{x}} = \frac{\sigma}{\sqrt{n}} = \frac{12}{\sqrt{9}} = 4 \text{ units}$$

The mean value of sampling distribution = 78 units. In other words, if we draw a series of samples consisting of 9 random observations from the parent population and calculate the sample means, then the sample means will be normally distributed around a mean value of 78 units with a standard error of 4 units.

Since the sampling distribution is normally distributed, we may make the following statements:

$$P[74 \leq \bar{x} \leq 82] = 0.682$$
$$P[70 \leq \bar{x} \leq 86] = 0.954$$

That is, the probability of obtaining a sample mean value $\bar{x}$ within an interval of one standard error on either side of the population mean value is 0.682, and an interval of two standard errors on either side of the mean value is 0.954.

Change of standard error with n. The standard error of the sampling distribution of means may be used to determine the number of observations required to achieve some degree of confidence in the mean of a set of observations.

Figure 4-16 illustrates the rate of change of the standard error of the sampling distribution of the sample means with sample size. The parent popula-

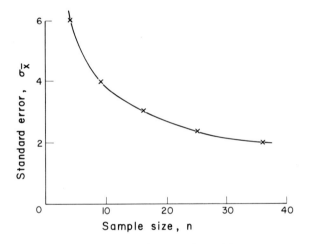

Fig. 4.16. Rate of change of $\sigma_{\bar{x}}$ with n.

tion used to illustrate this relationship is assumed to be normal with a standard deviation $\sigma = 12$ units. Note that for $n = 4$, $\sigma_{\bar{x}} = 6$, while for $n = 16$, $\sigma_{\bar{x}} = 3$. In other words, since the standard error is a function of $1/\sqrt{n}$, the standard error changes very slowly with increases in n.

Increasing the number of observations n is a poor way to improve the degree of confidence in a measurement. It is much more efficient to improve the intrinsic precision of a device or to replace the measurement device by one that is more precise.

Absence of unbiased errors. The concepts introduced above provide a means for condensing the information contained in repeated observations. With many measurement devices it may happen that repeated observations will not differ. In this case, the unbiased or random errors will be less than the finest scale readings of the particular device being used. Replicated measurements will serve to illustrate that the precision of the device is limited by its sensitivity.

4.9 Confidence Intervals

During the discussion of the measurement process it was pointed out that a measurement must be accompanied by a statement of the confidence that we have in a measurement. Just in the same way that standard units have been established for the communication of measurements, standard statistical techniques exist for calculating an estimate of the confidence characteristics of a measurement.

It is the purpose of this section to discuss the statistical techniques that exist for calculating the confidence levels associated with sample statistics, based upon both large and small numbers of observations.

The basic question. Confidence estimates are stated in the following form: What is the probability that the following statement is correct? "The interval $\bar{x} \pm \Delta$ will include X." That is, what is the probability that a certain range about our calculated sample mean will include the mean of the population. Since we are dealing with means, we shall need to use statistics dealing with means and not samples. We shall therefore have use for the standard error.

Figure 4.17 illustrates this concept graphically, and the statistical techniques described in this chapter will provide us with standard procedures for answering this question.

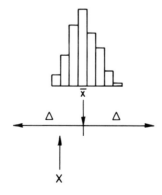

What is the probability that the following statement is correct ?
" The interval $\bar{x} \pm \Delta$ will include X."

Fig. 4.17. Concept of a confidence interval.

The interval $\bar{x} \pm \Delta$ is called a *confidence interval,* and the numbers which define the ends of the interval are called *confidence limits.*

Samples containing large numbers of observations. In Section 4.8, the sampling distributions of sample statistics that arise when random samples consisting of n observations are drawn from a parent population with known mean and standard deviation were discussed.

In particular, it was seen that the means, $\bar{x}$, are normally distributed and that this sampling distribution has a mean value equal to the population mean X and a standard error equal to $\sigma_{\bar{x}} = \sigma/\sqrt{n}$.

For a particular measurement problem we do not know the population parameters X and σ and we must estimate their magnitudes from information conveyed by the sample statistics $\bar{x}$ and s.

The estimation of the population mean X is the sample mean $\bar{x}$.

The estimate of the population standard deviation σ is given by the standard deviation s.

Since the sampling distribution of the sample mean is normal, we may make the following statement: The population mean value X lies within the interval $\bar{x} \pm 3\sigma/\sqrt{n}$, and the probability of this statement being correct is 0.997. The value 0.997 refers, of course, to the area under ± 3 standard deviations from the mean.

The interval is known as the *confidence interval* for the population mean X. The confidence interval is of fundamental importance to the measurement process since it expresses the reliability of our sample statistic $\bar{x}$. It should be recognized that the confidence interval is derived from the precision of the measurement device. This concept is valid only if the observations are free from biased errors. There is not necessarily a relationship between the confidence interval and accuracy.

For a particular set of observations, we may express any number of confidence intervals in the following form:

$$99.7\% \text{ Confidence interval: } \bar{x} \pm \frac{3\sigma}{\sqrt{n}}$$

$$95.0\% \text{ Confidence interval: } \bar{x} \pm \frac{1.96\sigma}{\sqrt{n}}$$

$$68.3\% \text{ Confidence interval: } \bar{x} \pm \frac{\sigma}{\sqrt{n}}$$

In other words, the more the width of the interval decreases, the lower is the probability that our assertion that the population mean value falls within the interval is correct. As an example, consider a sample with the following characteristics:

$$n = 36, \quad \bar{x} = 345 \text{ mm}, \quad \text{and} \quad s = 12 \text{ mm}$$

Standard error of sample distribution: $\sigma_{\bar{x}} = \dfrac{12}{\sqrt{36}} = 2 \text{ mm}$

99.7% Confidence interval: 345 mm ± 6 mm
95.0% Confidence interval: 345 mm ± 3.9 mm
68.3% Confidence interval: 345 mm ± 2 mm

The assertion that the population mean value falls within the relatively narrow interval of 343 – 347 mm has a probability of 0.683 of being correct. However, the assertion that the population mean value falls within a much broader interval of 339 – 351 mm has a probability of 0.997 of being correct.

In general, the confidence interval for the population mean X is given by $\bar{x} \pm \Delta$, where $\Delta = z \cdot \sigma/\sqrt{n}$ and z may be obtained from tables of the standard form of the normal distribution for various probability levels. Commonly used values are given in Table 4.6.

Table 4.6

Confidence level	99.9	99.9	95.0	90.0
z	3.30	2.57	1.96	1.65

It should be noted that the tables of the areas under the curve for various z values (viz. Table I.1) are for one side of the distribution only. The confidence levels in Table 4.6 represent the area contained within the band $\pm z$, or twice the area shown in the tables of z.

We have seen that for a given set of observations, we may calculate any number of confidence intervals that correspond to various probability levels. For a confidence interval to be meaningful in the communication of measurements, it would seem appropriate that we select a particular confidence interval for this purpose. Before suggesting a particular interval, we need to introduce a second term, which is, in fact, complementary to the confidence interval concept.

4.10 Levels of Significance

The level of significance refers to the probability that an assertion about the population mean value falling within a specific interval will be incorrect.

The key to the previous discussion was a probability statement about the assertion that the population mean value falls within a particular interval around the sample mean $\bar{x}$. For example, we noted that there is a 95% probability that the following statement is correct: The population mean falls within the interval $\bar{x} \pm 1.96\sigma/\sqrt{n}$.

Alternatively, we may state that there is a 5% (i.e., 100 − 95) probability that the above statement is incorrect. This probability is called the level of significance of a statistical test.

In other words, if we operated a measurement process a large number of times and obtained a series of samples containing n observations and from each of these samples drew the conclusion "The population mean value falls within the interval $\bar{x} \pm 1.96\sigma/\sqrt{n}$," then we could expect to be wrong in about 5% of the cases.

The term *significance* is used in a statistical sense of the word, and it means that the probability of an assertion, such as those made above arising from chance causes alone, is equal to the level of significance. There are three levels of significance which are frequently used:

$$5\% \text{ level}, \quad 1\% \text{ level}, \quad \text{and} \quad 0.1\% \text{ level}$$

The 1% level is usually referred to as significant.

The actual level of significance selected for a particular measurement process is a function of the value of various levels of accuracy. If small changes in accuracy are very valuable, then an observer might specify the 0.1% level. Normally intervals are expressed in terms of the 5% level of significance.

Determination of sample size. The relationship

$$\Delta = z \cdot \frac{\sigma}{\sqrt{n}}$$

may be used to estimate the sample size required to produce a sample average $\bar{x}$ of specified reliability.

For example, if we have a much-used measurement device which has a known precision represented by $\sigma = 12$ mm and we wish to establish a confidence interval for the mean X that is 5 mm on either side of the sample mean $\bar{x}$ at the 5% level of significance, then we may proceed in the following manner: For 5% level of significance $z = 1.96$

$$\Delta = \frac{z \cdot \sigma}{\sqrt{n}} \quad \text{or} \quad 5 = \frac{1.96 \cdot 12}{\sqrt{n}}$$

$$\sqrt{n} = 4.7 \quad \text{or} \quad n = 22$$

In other words, if we obtained samples containing, say, 25 observations, then we could make the following statement: The population mean value will fall within an interval 5 mm on either side of the mean value and there is a 95% probability that this assertion is correct.

Samples containing small numbers of observations. When we obtain small samples ($n <$ about 10) from a measurement device, the standard deviation s is not a reliable estimate of the population standard deviation σ. The problem of inferring something about the parent population from small samples was solved in the earlier part of this century by an Irish chemist who wrote under the pen name of Student. Instead of calculating the standard error of the sampling distribution of the sample mean from $\sigma/\sqrt{n}$ and then using this to estimate confidence intervals, he suggested that we use the expression

$$\Delta = \frac{t \cdot s}{\sqrt{n}} \tag{4.17}$$

where t represents a multiplier to account for the bias of small samples.

If samples consisting of n observations are drawn from a normal population with mean X, t may be computed from

$$t = \frac{\bar{x} - X}{s} \cdot \sqrt{n} \tag{4.18}$$

Student showed that the sampling distribution of the statistic t is given by

$$f(t) = \frac{Y_0}{\left(1 + \frac{t^2}{n-1}\right)^{n/2}}$$

$$= Y_0 \left(1 + \frac{t^2}{\nu}\right)^{(\nu+1)/2} \tag{4.19}$$

where Y_0 is a constant depending on n such that the total area under the curve is 1 and $\nu = n - 1$ is called the number of degrees of freedom.

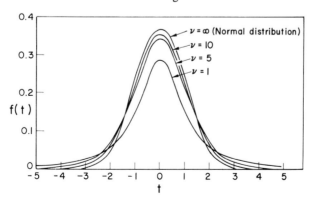

Fig. 4.18. t distribution for various degrees of freedom.

Figure 4.18 shows that t distribution for various degrees of freedom. Note that as the number of degrees of freedom increases, we approach the normal distribution, i.e., the distribution of z. Table I.2 shows the values of t for various levels of significance and for varying degrees of freedom. Table I.2 differs from Table I.1 in that instead of having one value of z for each significance level, we have one value of t for each value of the degrees of freedom.

Consider a sample of observations with the following characteristics:

$$n = 10, \quad \bar{x} = 17 \text{ units}, \quad \text{and} \quad s = 4$$
$$\text{For 5\% level:} \quad \nu = 9, \quad t = 2.262$$
$$\Delta = \frac{2.262 \times 4}{\sqrt{10}} = 2.864$$

$$\text{For 1\% level:} \quad \nu = 9, \quad t = 3.250$$
$$\Delta = \frac{3.250 \times 4}{\sqrt{10}} = 4.110$$

We may now make the following statement: There is a 95% probability that the assertion that the population mean falls within 17 ± 2.86 units is correct. We may also state that "There is a 99% probability that the assertion that the population mean falls within 17 ± 4.11 units is correct." An alternative statement is "The assertion that the population mean falls within 17 ± 2.86 units is significant at the 5% level."

Further determination of sample size. The relationship $\Delta = t \cdot s/n$ may also be used to estimate the sample size required to produce a sample average $\bar{x}$ of specified reliability.

Unlike the case for large samples discussed earlier, we cannot proceed directly with the calculation of n as t varies with n. We have to estimate n by trial and error. For example, for the measurement process just referred to above, assume that we would like to make the following statement: The population mean falls within a band of ± 3 units on either side of the sample mean $\bar{x}$ at the 1% level of significance:

$$\Delta = \frac{t \cdot 4}{\sqrt{n}} \quad \text{or} \quad t = 0.75 \times \sqrt{n}$$

for

$$n = 14, \quad t = 3.012, \quad \text{and} \quad t(\text{calculated}) = 2.81$$
$$n = 15, \quad t = 2.977, \quad \text{and} \quad t(\text{calculated}) = 2.90$$
$$n = 16, \quad t = 2.947, \quad \text{and} \quad t(\text{calculated}) = 3.00$$

In other words, for the above measurement device in which we have an estimate of its precision ($s = 4$), we would need to generate about 16 observations in order to make the confidence statement described above.

4.11 Exercises

4.1. (a) What is the maximum height of the standard form of the normal distribution?

(b) Why is the Bessel correction $\left(\sqrt{\dfrac{n}{n-1}}\right)$ necessary?

(c) Why is increasing sample size a poor way to increase precision?

(d) Why is it often desirable to test whether the results from experiments repeated many times follow the normal frequency distribution?

4.2. Data:

6.72	6.70	6.64	6.66
6.75	6.81	6.68	6.70
6.72	6.82	6.76	6.73
6.76	6.66	6.78	6.62
6.74	6.76	6.78	6.70
6.77	6.78	6.76	6.76
6.66	6.79	6.66	6.62
6.76	6.70	6.67	6.80

(a) Draw a histogram.

(b) Draw a relative frequency diagram.

(c) Draw a cumulative frequency diagram.

(d) Calculate the mean and standard deviation of the data (for s^*, first 10 data values only).

(e) Estimate the population mean and population standard deviation.

4.3. Evaluate the following for the data contained in Table E4.3:

(a) The modal and median salaries.

(b) The mean salary.

(c) The percentage of employees earning less than $115 per week but greater than the mean salary.

Table E4.3

Salary ($/week)	Number of employees
75.00–84.99	8
85.00–94.99	12
95.00–104.99	18
105.00–114.99	13
115.00–124.99	9
125.00–134.99	6
135.00–144.99	3
145.00–154.99	1
Sum	70

4.4. Construct a histogram, a relative frequency histogram, and a frequency distribution for the following observations of the weights of 40 male students:

138	164	150	132	144	125	149	157
146	158	140	147	136	148	152	144
168	126	138	176	163	119	154	165
146	173	142	147	135	153	140	135
161	145	135	142	150	156	145	128

Calculate the following:
(a) The modal weight.
(b) The mean weight.
(c) The standard deviation of weights.

Answer the following questions:
(a) What is the sum of the relative frequencies of all classes?
(b) How many students have weights within a band one standard deviation on either side of the mean weight?
(c) What proportion of the total number of students have weights within a band two standard deviations on either side of the mean weight?
(d) What proportion of students have weights greater than the mean weight plus three standard deviations?

4.5. The following results were obtained in measuring the length of a piece of equipment:

Length (cm)	2.72	2.73	2.74	2.75	2.76	2.77	2.78	2.79	2.80
Frequency	1	2	4	6	5	4	2	1	1

Identify modal and median length. Calculate the mean and the standard deviation of this set of measurements. How would you present the result of this series of measurements in condensed form?

4.6. Given the data 10.0, 10.5, 10.5, 11.0, 11.0, 11.0, 11.5, 11.5, 11.5, and 12.5, estimate the mean, the standard deviation, and the standard error of the population.

4.7. The weight of shipments arriving at a receiving department are normally distributed and have a mean weight of 200 lb and a standard deviation of 60 lb. What is the probability that 36 shipments received at random will exceed the specified capacity of 7500 lb of the freight elevator? What would be the probability if the standard deviation of the weights were 50 lb? What would the probability be if the mean weight was 250 lb and the standard deviation 55 lb?

4.8. The finished inside diameter of a piston ring is normally distributed with a mean of 4.50 in. and a standard deviation of 0.005 in. What is the probability of obtaining a diameter exceeding 4.51 in.?

4.9. The resistance of a foil strain gage is normally distributed with a mean of 120.0 ohms and a standard deviation of 0.4 ohms. The specification limits are 120.0 ± 0.15 ohms. What percentage of gages will be defective?

4.10. A ready-mix concrete plant produces concrete with a mean strength of 3300 psi and a standard deviation of 150 psi. What is the probability that the mean strength of *three* specimens obtained from this plant will have a strength between 3450 and 3600 psi? What is the probability that a single specimen will have a strength between 3450 and 3600 psi?

4.11. Plot a histogram for the following data from 100 tests of the shear strength of steel:

Strength (tons/sq in.)	No. of tests
14–15	1
15–16	0
16–17	3
17–18	16
18–19	18
19–20	24
20–21	21
21–22	14
22–23	2
23–24	0
24–25	1
Total	100

Plot the best-fitting normal curve on the histogram given that $s = 1.6$ tons per sq in. and $\bar{x} = 19.5$ tons per sq in.

4.12. Calculate the following probabilities:
(a) The probability of drawing the four aces from a deck of cards.
(b) The probability of drawing any four hearts from a deck of cards.
(c) The probability of drawing any heart from a deck of cards.

4.13. For a roulette wheel containing numbers 1–36 inclusive, a 0, and 00, calculate the following probabilities:
(a) The occurrence of 6.
(b) The occurrence of an odd number.
(c) The occurrence of any of the numbers 1, 2, or 3.

4.14. Calculate the following probabilities:
(a) The probability of drawing a three from a deck of cards.

(b) The probability of drawing a spade from a deck of cards.

(c) The probability of drawing 4 black balls in a row from a group of 20 balls (10 black and 10 white) if the drawn balls are *not* replaced.

4.15. (a) The heights of 528 first-year engineers have been observed and found to be normally distributed with a mean of 68.0 in. and a standard deviation of 4.0 in. Freshman year is divided into sections of 24 students on an essentially random basis (with respect to height). If the mean heights of each section were observed, what type of theoretical distribution would you expect these 22 means to follow? What would be the mean value of this sampling distribution of section mean heights? How many sections would you expect to have mean heights between

(1) 68.0 and 68.8 in.?

(2) 66.8 and 68.3 in.?

(3) Greater than 69.6 in.?

(4) Less than 66.4 in.?

(b) How many observations would be required to provide an estimate of the true height within ±1.0 in. at the

(1) 5% level of significance?

(2) 1% level of significance?

(3) 0.1% level of significance?

4.16. In Exercise 4.15, it was pointed out that the mean height of 528 first-year engineering students was observed to be 68.0 in. and the standard deviation of student heights was 4.0 in. Assume that this sample represents a random sample of the heights of all first-year engineering students in the country. Calculate for this sample mean the

(a) 99.9% confidence interval.

(b) 99.0% confidence interval.

(c) 93.0% confidence interval.

(d) 75.0% confidence interval.

Describe what these four confidence intervals mean, particularly in relation to each other.

4.17. (a) A sample of 10 observations of the diameter of a sphere gave a mean $\bar{x} = 4.38$ in. and a standard deviation of $s = 0.06$ in. Calculate the 95 and 99% confidence limits for the actual diameter.

(b) Calculate the above confidence limits using the methods of large sample theory. Compare the results of the two methods and explain why they are different.

4.18. Table E4.18 contains the characteristics of samples consisting of 36 observations. The mean and standard deviations of most of the samples are also given.

(a) Calculate the mean and standard deviations of those samples for which these statistics have not been calculated.

(b) What are the best estimates of the population parameters X and σ?

(c) Plot a graph which shows the sample mean value as the ordinate and the sample number along the abscissa. Plot the estimate of the population mean on this graph, showing it as a horizontal line.

(d) Plot a similar graph which shows the sample standard deviation as the ordinate and the sample number along the abscissa. Plot the estimate of the sample standard deviation on this graph, showing it as a horizontal line.

(e) Plot two lines on the graph of the sample mean which will enclose approximately 95% of all sample means generated by this process and which are based on 36 observations. Plot another two lines which will enclose approximately 99.7% of all sample means over the long run.

Table E4.18

Sample	Observations	Mean	Standard Deviation
1	22.53, 25.32, 26.81, 27.41, 28.93, 29.12, 30.33, 31.64, 32.83, 33.38, 33.59, 33.92, 34.51, 34.62, 34.81, 34.89, 35.13, 35.58, 35.65, 35.72, 35.78, 36.12, 36.32, 36.42, 36.82, 37.83, 37.14, 38.12, 38.93, 40.13, 41.02, 41.98, 43.01, 43.95, 45.21, 49.02		
2		34.6	4.73
3		33.2	3.73
4		34.8	4.55
5		33.4	4.00
6		33.9	4.30
7		34.4	4.98
8		33.0	5.30
9		32.8	3.29
10		34.8	3.77

4.12 Suggestions for Further Reading

Moroney, M. J., *Facts from Figures,* Penguin Books Ltd., Harmondsworth, Middlesex, England, 1956.

Neville, A. M., and Kennedy, J. B., *Basic Statistical Methods for Engineers and Scientists,* International Textbook Company, Scranton, Pa., 1964.

Wallis, W. A., and Roberts, H. V., *Statistics, A New Approach,* The Free Press, New York, 1956.

5

Treatment of Results

This chapter deals with the ways in which data taken directly from an experiment can be arranged and modified to be more useful and more clearly presented.

5.1 Basic Guidelines

The obtaining of *raw* data is only one of a series of steps to presentation of these data. The data must be recorded and *reduced.* This requires that the data be put in the proper units, used in some calculation, averaged, or otherwise transformed. It may be necessary to store the data for future use. Most importantly the data must be judged for accuracy and precision. In the end a method of presentation must be decided upon. Graphs, tables, formulas, etc., may be considered for this. We shall consider each of these stages in turn. Final presentation of data has been left to Chapter 6.

The first requirement for any experimental program is a journal or record of the step-by-step progress of the experiment. The history of science is filled with stories of how scientific discoveries were made by observing events which the experimenter could not explain. A careful search of his own logbook eventually turned up the reason and an explanation.

Everything should be recorded: date, temperature, instruments used (including serial numbers if many similar instruments are used in the same organization), dates and methods of calibration, etc. Surveyor's logbooks are excellent examples of the type of records which should be kept. Surveyor's logbooks, in fact, are legal documents and presentable as evidence in courts, which is also true of other properly kept and dated logbooks. Patent rights and claims to earlier discovery are often decided upon the basis of logbooks. In industrial organizations, test data, performance calculations, operating records, etc., are all permanently recorded. In many companies this information forms the basis of the design of new equipment. The proper keeping of a journal is the sign of an organized and experienced professional.

5.2 Data Reduction

For small experiments, data reduction seldom presents a problem. A slide rule or desk calculator and a table of calculations will reduce small amounts of data to the desired form. However, modern experimental work has many examples of problems with two additional requirements:

1. The data must be recorded quickly since the event measured occurs quickly.
2. The amount of data required is very large.

An example of the first case is given in Figure 5.1, which shows the variation of voltage across an extremely small resistor when a small raindrop hits it at high

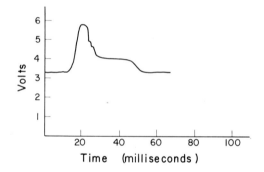

Fig. 5.1. Variation of voltage across a small resistor when hit with a small water droplet.

velocity. The record of the raw data is, in fact, the photograph of the face of the oscilloscope (from which this figure was taken).

As an example* of the second problem, a carefully planned study of vibrations in a radar aerial while tracking an airplane was studied by mounting a camera with telescope and cross hairs on the aerial and following an airplane. The program required 162,000 frames of movie film to be read by eye for a probable error up to 10% in the required final result. The reason for the large size of the error and the large number of measurements taken is related to the fact that the frequency of vibration of the aerial varies over a wide range and each frequency must be studied. A more detailed analysis of this problem requires a sophisticated mathematical approach.

*R. B. Blackman, and J. W. Tukey, *The Measurement of Power Spectra*, Dover Publications, Inc., New York, 1958, p. 58.

The computer is, of course, the tool for handling large data reduction problems. The problem of the radar aerial required 24 hr of large-computer time to process the data from the film frames. Computers may handle experimental data in many ways:

1. They can simply transform the data by putting them into the proper units and the like. Small and large computers are available for this work depending on the degree of difficulty of the analysis or the quantity of data.

2. The computer can be *on line.* The signal from the experiment or test can be converted to an electrical signal (if it is not in this form already). This electrical signal can be directly "plugged in" to the computer, which can then print the results in a suitable form. The computer can even plot the results of an experiment. Figure 5.2 shows the velocities of air in a fluidic valve. Each curve on the paper has been calculated from 161 measured data points. *Every* line and letter on this page has been produced by the computer.

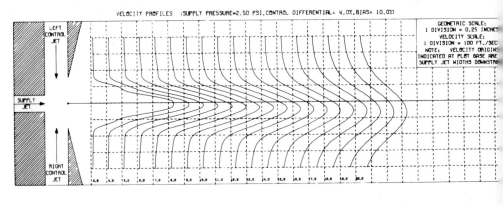

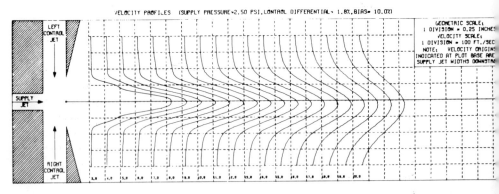

Fig. 5.2. Computer-plotted graph.

3. The computer can be used to run an experiment as well as analyze data. When controlled by a computer a temperature probe can, for example, be arranged to hunt for a maximum temperature or a constant temperature (thus producing lines of constant temperature by its path).

It should not be necessary to say so but *sophistication should not be pursued as an end in itself. It is expensive in terms of both time and money.* The appeal of sophisticated hardware is nearly impossible for some people to resist. It has a valuable purpose, but not everywhere. Let your project drive you to sophistication, not vice versa.

Here is an example of the simplicity of approach of a good observer (Sir Geoffrey Taylor, a famous British physicist). He is reporting events from London in 1915.

> On one occasion when I had risked going home for the night I was woken at my home in London by gunfire. I saw the whole sky lit up and realized that something spectacular was happening on the other side of the house. I ran across a passage and saw an airship which turned out to be a Schutte-Lanz coming down vertically in flames. I watched it till it came down behind a chimney of a house 150 yards away, whose garden backed onto ours. I noticed that the ship appeared just the same length as the chimney pot. When daylight came I got out my sextant and found that the chimney subtended about 35 minutes of arc and its compass bearing was NNE. I knew that airships were about 600 feet long and 600 feet subtends 35 minutes of arc at about 11 miles. I laid off this compass bearing and distance on a map and found that the nearest village with a railway station to this position was called Cuddley. I bicycled to King's Cross Station and took the first train to Cuddley at about 5:30 a.m. and found that several local inhabitants had got up early and were making their way in a certain direction. I followed and found the remains of the airship concentrated in an area about equal to its cross-section. The bodies of the crew were still scattered around when I got there.*

5.3 Data Storage

Original data should not be thrown away. It is common for future uses to be found for data which have been obtained. In the study of fluid flow in a pipe, a piece of analysis in 1959 predicted that data which had been accepted as correct since 1928 were, in fact, likely to be in error. A study of the original 1928 data proved that an error had been introduced in calculating the final result. Hence, incorrect final data had been used by engineers in this field for 31 years as a guide to design. Failure to keep the original form of the data would have required new

*Address of Professor Sir Geoffrey Taylor to the Third Canadian Congress of Applied Mechanics, Calgary, Canada, May 1971. Quoted with permission.

experiments and the old data would still have caused confusion in its incorrect form.

Industries and government have typically kept data and original engineering drawings in vaults and fireproof cabinets. The modern trend, however, is to use computer or microfilm methods to store information. In this respect, data handling and engineering drawing storage are much ahead of libraries in their storage technology. The usual methods for computers are magnetic tape and punched cards.

5.4 Least Squares or Regression

In Chapter 4 we were involved in measurements of a single quantity a number of times. We are now going to consider measuring a function over a range. For example, consider measuring the diameter of a tree on, say, the first of November every year for 6 yrs. The resulting graph could be something like that in Figure 5.3. The scatter of the measurements is due to errors in the height at which we measure the tree in different years, the pressure of our measuring device on the bark, and damage to the tree during a previous year. The height would be difficult to set each year because the soil level can change from year to year.

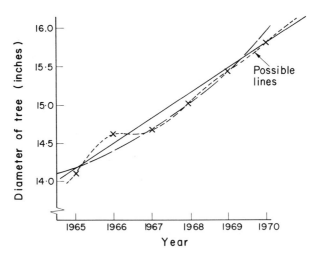

Fig. 5.3. Diameter of a tree measured on November 1 every year.

We begin by asking the question "How can we best draw a line through these points?" It is the purpose of this section to answer this question. Several lines are possible, as we can see from Figure 5.3.

Two points should be noticed about our problem. First, it is reasonable to assume that nearly all the errors are in the diameter measurement. It is not difficult to arrange to measure at the same date every year, and so we can assume that the error in the determination of the time (the horizontal axis) is small or effectively zero. Second, we should note that the simplest functional variation is, of course, a straight line. Any linear variation of a function is much easier to handle than a curved variation. For want of better information, let us assume that we wish to draw a straight line through the points of Figure 5.3. Later in this section we shall discuss methods of overcoming this severe restriction. We can mention now, though, that it will always be necessary in using the method we are going to outline to assume some functional form (polynomial, exponential, etc.) for the curve we wish to plot.

We can assume, then, that we wish to draw a straight line through some data and that errors are in the vertical direction on the graph. If we draw a line $y = f(x)$ through the data as in Figure 5.4, the discrepancies between the points

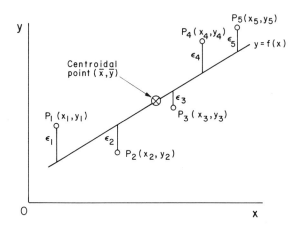

Fig. 5.4. Straight line showing residuals between data points, P_i, and line.

and the drawn line may be drawn as ϵ_1, ϵ_2, ϵ_3, etc. These quantities are called *residuals*. The "best" line is assumed to be that which makes the sum of the squares of the residuals a minimum. The arithmetic mean discussed in Section 4.3 was that quantity which made the sum of the deviations from it a minimum and so we see that this line has a partial analogy to the arithmetic mean of a sample. To find the line it is necessary only to add the squares of the residuals, arrange for them to be a minimum, and find what conditions this sets upon the line. The finding of minima is a simple exercise in calculus, and we shall now proceed to do this.

Let the equation of the line be

$$y = a + bx \tag{5.1}$$

It is now obvious that our final result must enable us to calculate a and b.

For a point x_i, y_i on the line, it is further obvious that

$$y_i - (a + bx_i) = 0$$

However, for a point not on the line this will not be true and the difference between y_i (actual) and $a + bx_i$ on the line, evaluated at x_i, will be equal to the residual ϵ_i.

Hence, for an experimental point

$$y_i - (a + bx_i) = \epsilon_i \tag{5.2}$$

The sum of the squares of the residuals is $\Sigma\, \epsilon_i^2$, and

$$\Sigma\, \epsilon_i^2 = \Sigma\, [y_i - (a + bx_i)]^2 \tag{5.3}$$

We must set this quantity at a minimum by determination of the two variables a and b. Hence, we wish to set

$$\frac{\partial\, \Sigma\, \epsilon_i^2}{\partial a} = 0$$

and

$$\frac{\partial\, \Sigma\, \epsilon_i^2}{\partial b} = 0$$

Differentiating equation (5.3) first with respect to a and then to b, we obtain

$$\frac{\partial\, \Sigma\, \epsilon_i^2}{\partial a} = -2\, \Sigma\, [y_i - (a + bx_i)]$$

$$\frac{\partial\, \Sigma\, \epsilon_i^2}{\partial b} = -2\, \Sigma\, x_i\, [y_i - (a + bx_i)]$$

That is,

$$\Sigma\, [y_i - (a + bx_i)] = 0 \tag{5.4}$$

and

$$\Sigma \{x_i [y_i - (a + bx_i)]\} = 0 \tag{5.5}$$

Since all x_i and y_i are known, equations (5.4) and (5.5) represent two equations in two unknowns a and b so that a and b may be solved for.

Consider equation (5.4) first:

$$\Sigma [y_i - (a + bx_i)] = \Sigma y_i - \Sigma a - b \Sigma x_i = 0$$

Since a is a constant, $\Sigma a = na$. For convenience we now drop the subscripts but remember their meaning. Dividing through by n, we obtain

$$\frac{\Sigma y}{n} - a - \frac{b \Sigma x}{n} = 0 \tag{5.6}$$

But

$$\frac{\Sigma y}{n} = \bar{y} \quad \text{and} \quad \frac{\Sigma x}{n} = \bar{x}$$

$$\bar{y} = a + b\bar{x} \tag{5.7}$$

Equation (5.7) then tells us that the line we require passes through the *centroidal point* of the data, which is the point $\bar{x}$, $\bar{y}$ on the graph.

Equation (5.5) now gives

$$\Sigma [x_i y_i - ax_i - bx_i^2]$$

or

$$\Sigma xy - a \Sigma x - b \Sigma x^2 \tag{5.8}$$

We may now simply solve equations (5.6) and (5.8) to obtain

$$a = \frac{\Sigma x^2 \, \Sigma y - \Sigma x \, \Sigma xy}{n \, \Sigma x^2 - (\Sigma x)^2} \tag{5.9}$$

and

$$b = \frac{n \, \Sigma xy - \Sigma x \, \Sigma y}{n \, \Sigma x^2 - (\Sigma x)^2} \tag{5.10}$$

The values we compute for a and b according to equations (5.9) and (5.10) when placed in $y = a + bx$ will give the line where the sum of the residuals $\Sigma\ e_i^2$ is a minimum. The resulting equation is called the *regression of y upon x. y* has been regressed upon x since the errors have been assumed to be in y, not in x.

If the errors had been in x and not y, it would be possible to perform the regression of x upon y. Then if

$$x = \alpha + \beta y$$

α and β are given by the same reasoning as we obtained a and b. The resulting formulas for α and β are

$$\alpha = \frac{\Sigma y^2\ \Sigma x - \Sigma y\ \Sigma xy}{n\ \Sigma y^2 - (\Sigma y)^2} \qquad (5.11)$$

$$\beta = \frac{n\ \Sigma xy - \Sigma x\ \Sigma y}{n\ \Sigma y^2 - (\Sigma y)^2} \qquad (5.12)$$

Note that the same lines do not result but that both lines pass through $\bar{x}$, $\bar{y}$.

Example

Given the data

$$x = 1.0, \quad 2.0, \quad 4.0, \quad 4.0, \quad 5.0, \quad 6.0$$
$$y = 2.02, \quad 4.00, \quad 5.98, \quad 7.90, \quad 10.10, \quad 12.05$$

find the regression line of y upon x. The best method of proceeding in a calculation of this type is to work from a table of data:

x	x^2	y	xy
1.0	1.0	2.02	2.02
2.0	4.0	4.00	8.00
3.0	9.0	5.98	17.94
4.0	16.0	7.90	31.60
5.0	25.0	10.10	50.50
6.0	36.0	12.05	72.30
21.0	91.0	42.05	182.36
Σx	Σx^2	Σy	Σxy

$$a = \frac{\Sigma x^2\ \Sigma y - \Sigma x\ \Sigma xy}{n\ \Sigma x^2 - (\Sigma x)^2} = \frac{91.0 \times 42.05 - 21.0 \times 182.36}{6 \times 91.0 - (21.0)^2}$$

$$= -0.0287$$

$$b = \frac{n \, \Sigma \, xy - \Sigma \, x \, \Sigma \, y}{n \, \Sigma \, x^2 - (\Sigma \, x^2)} = \frac{6 \times 182.36 - 21.0 \times 42.05}{6 \times 91.0 - (21.0)^2}$$

$$= +2.01$$

The result is plotted in Figure 5.5.

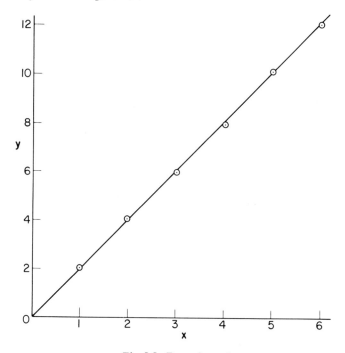

Fig. 5.5. Example results.

The problem of fitting curves other than straight lines to a set of data may be treated in several ways. For the simpler cases we simply change the form of the graph which we plot. For example, if we wish to fit $y = a + b/x$ to a set of data, we may plot y against $1/x$; then a and b are determined as before. For the curve $y = ax^b$ we may write that

$$\log y = \log a + b \log x$$

and plot $\log y$ against $\log x$.

There are a number of problems which arise in regression analysis which we have not dealt with. The study of regression can be quite difficult if we ask ourselves such questions as "What if the errors are significant for both x and y? What if the errors become larger as the value measured gets larger?" In this case, the data will become more scattered as we move to the right on our graph. An example is given in Figure 5.6.

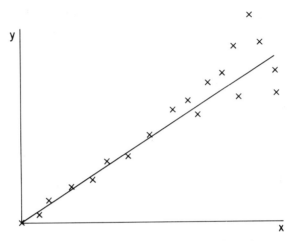

Fig. 5.6. Increased scatter for larger values of x and y.

There is one question which we can answer with our present tools. Having obtained the line which is required, how good a fit is this to the data? For example, there is definitely a difference in the cases given in Figure 5.7(a) and (b).

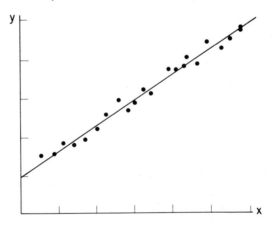

Fig. 5.7. (a) Good correlation.

The same straight line fits both sets of data. We distinguish between the two cases by saying that the data in Figure 5.7(a) are better correlated than those in Figure 5.7(b). The difference in the two graphs is, of course, that the data are more dispersed in Figure 5.7(b). We are therefore looking for a measure of dispersion about the line. This quantity will bear certain resemblances to the standard deviation of a sample since the purpose is similar.

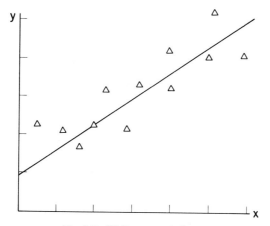

Fig. 5.7. (b) Poor correlation.

The quantity used to define the dispersion is called the product moment correlation coefficient or, more simply, the correlation coefficient r. The correlation coefficient is given by

$$r = \frac{n \Sigma xy - \Sigma x \, \Sigma y}{[n \Sigma x^2 - (\Sigma x)^2][n \Sigma y^2 - (\Sigma y)^2]} \tag{5.13}$$

By means of algebraic manipulation we can cast this equation in two different forms which will help us to understand its significance.

Equation (5.13) may be written as

$$r = \frac{(1/n) \Sigma (x - \bar{x})(y - \bar{y})}{\sqrt{[\Sigma (x - \bar{x})^2/n] \cdot [\Sigma (y - \bar{y})^2/n]}} \tag{5.14}$$

The sign of r is always taken as positive. The product term $(1/n) \Sigma (x - \bar{x})(y - \bar{y})$ in the numerator of equation (5.14) is called the covariance of x and y. The two terms $\Sigma (x - \bar{x})^2/n$ and $\Sigma (y - \bar{y})^2/n$ appearing under the root are the sample variances of x and y. We may then say that the correlation coefficient r is the covariance of x and y divided by the square root of the product of the variances of x and y.

The covariance, in turn, may be understood in a physical sense by the following argument. Select a value of x. Make repeated measurements of y for this value of x. If there is a strong functional relationship between x and y, the value of y will repeat itself. In this case the value of $\Sigma (x - \bar{x})(y - \bar{y})$ will continue to get larger the more values we take. Stating this another way, a strong correlation will give a significant nonzero value to $(1/n) \Sigma (x - \bar{x})(y - \bar{y})$ for a specific value of x. If there is no correlation, then $y - \bar{y}$ can be alternately positive or negative. The average of this would tend toward zero if the correlation were so poor that $y - \bar{y}$ had an equal chance of being positive and negative while $x - \bar{x}$ retained a constant value. We see, then, that for a good correlation the covariance will be large. For a poor correlation, it will be small or even zero. The presence of the two variances in the denominator of the equation can be considered as a method of nondimensionalizing or normalizing the quantity, Nondimensionalizing and normalizing are discussed in Section 5.5.

Another way of considering the equation for r is to show that it can be written as

$$r = \sqrt{b \cdot \beta} \qquad (5.15)$$

This is proved by algebraically manipulating the products of the right-hand sides of the equations for b and β [equations (5.10) and (5.12)].

If there is no correlation between x and y, the data will look like that in Figure 5.8. The regression of y upon x will give a horizontal line (line I), since this line will minimize the sum of the vertical residuals. The regression of x upon y will produce a vertical line (line II), to minimize the sum of the horizontal residuals. The result of this is that b (the slope of line I with respect to the x axis) is zero and β (the slope of line II with respect to the y axis) is also zero. Hence,

$$r = \sqrt{b \cdot \beta} = \sqrt{0 \cdot 0} = 0$$

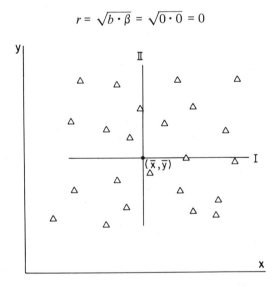

Fig. 5.8. Zero correlation.

If the correlation is perfect, then the two regressions will produce the same line through the data, as in Figure 5.9. Then $\beta = 1/b$ and

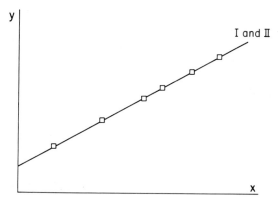

Fig. 5.9. Perfect (= 1.0) correlation.

$$r = \sqrt{b \cdot \frac{1}{b}} = 1.0$$

Therefore, as before, a perfect correlation will produce $r = 1$ and no correlation will produce $r = 0$.

The correlation is best calculated by means of a table similar to that used in regression. As before, a desk calculator or computer is a useful aid in performing these calculations.

5.5 Dimensional Analysis

In this section we shall be considering dimensions and units. A dimension is the type of quantity being measured. Lengths, times, velocities, temperatures, etc., are dimensions. Dimensions are expressed in units. Feet, meters, chains, light years, etc., are units of length.

The consideration of dimensions and units is fundamental to any study of the physical world. The purpose of evaluating dimensions is to compare one thing to another. A 2-ft ruler is simply a way of saying that the ruler is twice as long as a 1-ft ruler. That is, the longer ruler has been compared to a shorter one and found to be longer in the ratio 2:1. This simple concept can lead to some quite sophisticated ideas.

The first question which naturally arises is, if we wish to measure something, to what do we compare it? The defined units, feet, degrees Fahrenheit, seconds, etc., are usually used. The use of these defined quantities may, however, not be appropriate to the problem. For example, the problem may look needlessly complicated if we use inappropriate units. In Section 4.7, we discussed converting a value in a sample of z units in order to express that value in terms of the number of standard deviations from the mean at which it lies. By expressing a value in z units, it became a simple matter to obtain information about the proportion of values which might be larger or smaller than this value and about the *relative* closeness of the value to the mean. By comparing the value to the standard deviation we are using the dimensions which are most natural to the problem. To illustrate the methods of dimensional analysis we shall consider a number of examples.

Example 1

Figure 5.10 shows the variation of position of the tip of a vibrating cantilever beam as a function of time. The vertical axis is in inch units and the hori-

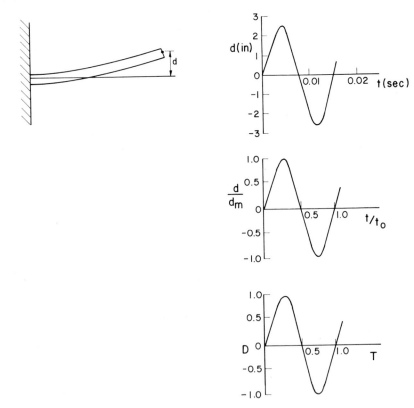

Fig. 5.10. Forms of equation for vibrating cantilever with no damping.

zontal axis is in seconds. We may write the equation for this curve by considering the form of the graph. It may be written as

$$d = 2.56 \sin \frac{2\pi t}{0.015} \qquad (5.16)$$

The presence of the values 2.56 and 0.015 in equation (5.16) do not make the equation impossible to handle, but the physical meaning of the terms is obscure. For example, the value of d at $t = 0.00375$ sec is

$$d = 2.56 \sin \frac{0.00375 \times 2\pi}{0.015}$$

$$= 2.56 \sin \frac{\pi}{2} = 2.56 \text{ in.}$$

The equation may be transformed to make its algebraic form simpler. If we define d_m as the maximum value of displacement (2.56 in.) and t_0 as the time (0.015 sec), then

$$\frac{d}{d_m} = \sin 2\pi \frac{t}{t_0}$$

and if we further define $d/d_m = D$ and $t/t_0 = T$, then the equation may be written as

$$D = \sin 2\pi T \qquad (5.17)$$

and the additional statements required are

$$D = \frac{d}{d_m}, \qquad T = \frac{t}{t_0}$$

For our particular problem $d_m = 2.56$ in. and $t_0 = 0.015$ sec.

There are several reasons we should prefer equation (5.17) to equation (5.16). First, equation (5.17) is dimensionless. The lengths and times appear as ratios only. Equation (5.17) therefore applies equally well for lengths and times expressed in any units. Second, equation (5.17) is algebraically simpler. If this equation must be used in other work, the second form is preferable. Third, it is easier to give a physical description of the phenomenon using equation (5.17). We previously used equation (5.16) to evaluate d at $t = 0.00375$; this value of T corresponds to $t/t_0 = T = 0.25$. Since $T = 1$ represents the end of a single complete cycle, we immediately see that $T = 0.25$ is one fourth of the way along the cycle and therefore $D = 1$ and $d = d_m$. The physical meaning of D is also clear. It is the proportion of the maximum displacement reached for the corresponding proportion of the total cycle, T. Transforming into units of the type of D and T is called *normalization*. The purpose of normalization is to force the range of a variable into a restricted range. For example $-1 \leq D \leq 1$. The normalized form of the error curve has an area equal to 1.0.

The purpose of the previous example has been to illustrate that the units which most properly describe a phenomenon are not necessarily or even usually the defined units such as feet and inches. Further, the best units to use are those which are inherent to the problem. Another way of saying this is that we usually compare one part of a phenomenon to other parts of the same phenomenon.

In many of the physical sciences, it is a fundamental problem to find a unit with which other measurements should be compared. In other words, we look for the *scale* which is most important to the problem. Some examples will make this clear.

Example 2

The time scale or period in this case of the vibrating beam problem was t_0. All other times are compared to this. The time $t_1 = t_0/2$ could also have been chosen. What is useful is to have this scale of time to compare with other times. The length scale of the problem which we chose was d_m.

Example 3

Figure 5.11 shows a graph of the velocity of water in a circular pipe plotted against the distance across the pipe. If the velocity is fairly low, a plot of this sort

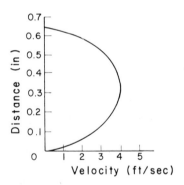

Fig. 5.11. Velocity distribution for flow in a circular pipe.

will result in a similar graph for pipes of other diameters and for flows which are larger or smaller. If we choose as a length scale the pipe radius and as a velocity scale the maximum velocity of water in the pipe (which occurs in the center), we may replot Figure 5.11 as in Figure 5.12. The advantage is that now we may hope that our new graph of V/V_{max} versus y/r is universal, that is, that it applies for

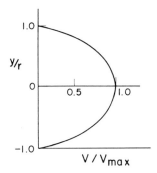

Fig. 5.12. Normalized velocity distribution for a circular pipe.

all pipe sizes and all values of V_{max}. If that were so, then we would only need to find V_{max} in a pipe of known radius in order to be able to use Figure 5.12 to tell us the velocity everywhere in the pipe. In fact, Figure 5.12 represents a universal curve of velocities only in a restricted range of velocities of water. For high velocities, this curve will give quite incorrect answers. The case of flow of a liquid in a pipe represents a case where the choice of scales (in this case length scale) represents a very difficult problem.

Example 4

Radioactive nuclei decay to other forms according to an exponential law. The general formula for this type of charge is

$$\frac{\eta}{\eta_0} = e^{-0.693\, t/\tau}$$

where η is the number of radioactive nuclei at some time t and η_0 is the number of nuclei which were radioactive at time $t = 0$. τ is the half life of the element expressed as a time. The time τ is that time at which exactly half of the nuclei have decayed. We may prove this by putting $t = \tau$ in the equation. Then

$$\frac{\eta}{\eta_0} = e^{-0.693} = 0.50$$

The graph of η versus t is shown in Figure 5.13. (Note that this is simply Figure 2.3 upside down.) This curve is universal for all radioactive elements because, first, it is dimensionless and, second, it has been found experimentally that this curve gives a universal description for all elements. The value of τ varies depending on the type of nucleus being considered. We can see from Figure 5.13 that a time scale based upon the total time for all nuclei to decay is not possible since that time is infinity.

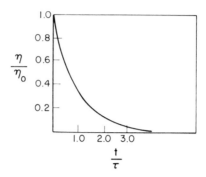

Fig. 5.13. Universal radioactive decay curve.

The examples we have treated so far have been comparisons of one measurement with another of the same sort. That is, we have divided lengths by lengths and times by times, etc., to produce dimensionless ratios which have (usually) some easily recognized interpretations. It is possible to produce dimensionless equations and, therefore, possibly more widely applicable equations by grouping quantities. The simplest examples of this are trivial. Newton's second law

$$F = kma$$

may be made dimensionless by dividing both sides by ma, i.e.,

$$\frac{F}{ma} = k$$

This form does not really give us a more useful form to work with, although it does hint at a rephrasing of the law as follows: The ratio of a force to the product of the mass upon which it acts and the resulting acceleration is constant.

Since all equations of whatever form must be dimensionally consistent, all equations can be nondimensionalized by dividing by one of the terms. By dimensionally consistent we mean that each term must be in the same units. For example, consider $y = a + bx + cx^2$. If y is in cubic feet, bx and cx^2 must also be in cubic feet. In addition, if the equation is correct for cubic feet, it must also be correct for other volume units, for example, cubic meters. (This is not always true for purely empirical equations.) We can make use of this fact using a method called dimensional analysis. If we do not know the form of an equation, we can consider the consequences of knowing that the unknown equation must be available in a nondimensional form. Consider the following problem: If we have a pipe and we wish to know how the friction affects the average flow through the pipe we may reason as follows: The friction coefficient is required. This may be considered as the ratio of the friction force of the walls to the force which could be exerted by the fluid flow. Hence, this coefficient is dimensionless. On what does the friction depend? Let us say it depends on the average velocity of the flow (U), the viscosity of the flow (μ), the diameter of the pipe (D), and the density of the fluid (ρ).

We can then write for the friction coefficient (F)

$$F = \text{function of } (U, D, \rho, \mu) \qquad (5.18)$$

The dimensions of these quantities are

$$D = \text{Length}$$
$$U = \text{Length/Time}$$
$$\rho = \text{Mass/(Length)}^3$$
$$\mu = \text{Mass/Length} \times \text{Time}$$

The grouping of the variables, U, D, ρ, μ, must be such that they are dimensionless. A simple arrangement of these quantities which is dimensionless is

$$\frac{UD\rho}{\mu}$$

We can now write equation (5.18) as

$$F = f\left(\frac{UD\rho}{\mu}\right) \qquad (5.19)$$

The consequence of this analysis is that if the quantity $UD\rho/\mu$ does not change, then F will not change. If we double the velocity and halve the diameter, the value of F will remain constant. This is new information. Further, if we perform an experiment, we need to vary only the product $UD\rho/\mu$, and we may do so by changing whichever of U, D, ρ, and μ is the most easily varied. The results should then be plotted on a graph of F versus $UD\rho/\mu$, and the results should apply to cases not measured but of similar values of $UD\rho/\mu$.

It should be noted that we have assumed that F depends on the stated quantities. In fact, it might depend on other things also. The resulting equation (5.19), like all equations which represent hypotheses, must be checked against experiment to see if it represents the data well.

In fluid mechanics, the grouping of dimensions, $UD\rho/\mu$, is of considerable importance and is given the name Reynolds number, after Osborne Reynolds, who discovered that a number of important fluid variables depend on it.

There are many other useful dimensionless quantities spread throughout the sciences. The simplest is probably π (the dimensionless ratio of circumference to radius of a circle). Others are friction coefficients, f numbers of lenses (the focal length divided by the lens diameter), and the z units of statistics.

Let us consider another example of the use of dimensions: The data of Figure 5.14 were produced by a student who was measuring the life of flashlight

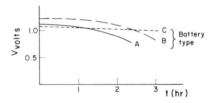

Fig. 5.14. Voltage versus time.

batteries when connected to a standard flashlight bulb. We can see by considering the graph that two different effects have been found. The various batteries have been found to have different voltages when first connected and they also decay in voltage at different rates. Can we arrange these data in a different way to make the conclusions clear? The purpose to which we put these data will determine our method. Suppose that we wish to know at what time the voltage will drop to 90% of its original value. To show this we may plot the data as in Figure 5.15,

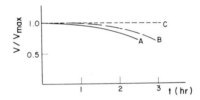

Fig. 5.15. Voltage divided by maximum voltage V_{max} versus time.

with the ordinate plotted as V/V_{max} and the abscissa as time. A table of V_{max} is also necessary, but this form of the graph makes the relative quality of the batteries quite clear as measured against our criterion of 10% drop from original value. There are numerous other ways to plot these data. Each one may illustrate a different way of considering the data. Figure 5.16 shows the data plotted as V

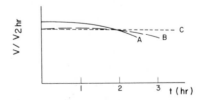

Fig. 5.16. Voltage divided by voltage after 2 hr, V_{2hr}, versus time.

divided by the voltage after 2 hr against time. From this graph we can consider the relative quality of the batteries if we wish to replace them after 2 hr of use.

Figure 5.17 shows the voltage divided by 1.10 v. From this graph we can consider the relative usefulness of each battery if they will no longer perform their

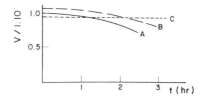

Fig. 5.17. Voltage divided by 1.10v versus time.

function below 1.10 v. We can see then from this simple example that the type of presentation and the form of nondimensional grouping will depend very much on the objectives of the test presented.

5.6 Units

The units which are defined according to some external standard are usually used in performing raw measurements and are also frequently the units in which data are presented. These units are simply those which are of a size range and popularity as to be suitable comparisons for a wide range of applications. If there is no obvious unit internal to a problem, then it is sufficient to report in these units. It would be of very little use to report room dimensions in terms of proportion of building lot width to a prospective house buyer. For modular building and prefabricated buildings these ratios would have value. The defined units are also the units in which most raw measurements are taken. Most instruments available for measurement are calibrated in these units in order to make them as versatile as possible. All these defined units are in turn defined in terms of four primary or basic dimensions: mass (M), length (L), time (t), and temperature (θ). The defined units are the kilogram, meter, second, and degree Kelvin. All other units are (as of 1969) defined in terms of these fundamental standard units. The definitions are given in Table 5.1.

In the theoretical sense, length, mass, and time alone should be sufficient to define all units since temperature could be defined in terms of energy of molecules. In practice, however, it is not possible to do this so the extra "fundamental" unit is required.

The numerical constants appearing in Table 5.1 are exact numbers. They are not rounded off, and there is no experimental error in them.

All other quantities, such as force, velocity, volts, charge, and power, are defined in terms of these fundamental quantities and a few defined constants such as g_c, the gravitational acceleration, and R, the universal gas constant. For example, velocity is defined as L/T. For conversions from one set of units to another there are various tables available in engineering and scientific handbooks.

Table 5.1.   Definition of Units of Primary Quantities (as of 1969)

Unit	Definition
Kilogram	The mass of a cylinder of platinum-iridium alloy kept at the International Bureau of Weights and Measures at Sèvres, France.
Meter	1,650,763.73 wavelengths in vacuum of the radiation corresponding to the transition between energy levels of $2p_{10}$ and $5d_5$ of the atom Krypton 86
International second	9,192,631,770 periods of the radiation corresponding to the transition between the two hyperfure levels $F = 4$, $m_F = 0$ and $F = 3$, $m_F = 0$ of the fundamental state $2S_{1/2}$ of the atoms of Celsium 133 undisturbed by external fields
Degree Kelvin (Celsius)	1/273.16 of the temperature of the system consisting of liquid water, water vapor, and ice in equilibrium (0.01°C)
Pound mass	0.45359237 kg
Foot	0.3048 m (1 in. = 2.54 cm)
Degree Rankine	1/1.8°K

It should be pointed out that the foot-pound system of engineers in North America is no longer a fundamental system in the sense that all units in this system are now defined in terms of the metric system of Table 5.1.

It is likely that in the next 5 yrs only Canada and the United States will be using the foot-pound system, and in these countries they will be used only in certain engineering and consumer fields. Scientific and most medical practices now are based upon the metric system in North America. Britain will have converted to metric measurements by 1975. It seems, however, that they will drive on the left-hand side of the road until the island drops into the sea.

The most modern and most commonly used type of metric system, of which there are several, is Système Internationale d'Unités or S.I. system. The S.I. system is made up of some nonmetric units. For example, the second is not a true metric unit in the sense that powers of 10 are not used to form units larger than seconds. The S.I. system has been adopted, or is in the process of being adopted, in, for example, Germany, Switzerland, the Netherlands, Sweden, Hungary, South Africa, and Great Britain. All attempts to recommend adoption of a metric system in the United States and Canada in the past few years have recommended the S.I. system. The other types of metric system differ mainly in the acceptable subunits which are employed. For example, some systems allow centimeters, while others would use only meters and millimeters in the same range.

5.7 Propagation of Errors and Approximate Numbers

Let us assume that we have measured a mass and an acceleration and that we wish to calculate the resulting force. Both the mass and the acceleration will have errors. How will these affect the error in the force? In other words, how will errors and uncertainties propagate through a calculation? To calculate this we make use of the chain rule of differential calculus.

If y is a function of x_1, x_2, x_3, etc.,

$$y = f(x_1, x_2, x_3, \ldots) \tag{5.20}$$

Then the chain rule states that

$$\delta y = \frac{\partial y}{\partial x_1} \delta x_1 + \frac{\partial y}{\partial x_2} \delta x_2 + \frac{\partial y}{\partial x_3} \delta x_3 + \cdots \tag{5.21}$$

If the increment of y is considered as the error in y and the increments δx_1, δx_2, δx_3, etc. are considered as the errors in x_1, x_2, x_3, etc., then equation (5.21) gives us a formula for δy. To obtain the maximum possible value of δy it is necessary to add the errors. Therefore,

$$\delta y = \left| \frac{\partial y}{\partial x_1} \delta x_1 \right| + \left| \frac{\partial y}{\partial x_2} \delta x_2 \right| + \left| \frac{\partial y}{\partial x_3} \delta x_3 \right| + \cdots \tag{5.22}$$

This would give the worst case. It is also possible to use the δxs corresponding to particular confidence or significance levels.

While this method of calculating errors is often used, it is pessimistic since the worst case is considered. In other words, if significance levels are used for the δx_i in equation (5.22), the resulting significance level of δy will be much less. Again, the resulting error δy will be pessimistic and large. It will therefore represent a lower significance level.

A better (but not exact) formula which gives *approximately* the same significance level for δy as the δx_is is

$$\delta y = \left[\left(\frac{\partial y}{\partial x_1} \delta x_1 \right)^2 + \left(\frac{\partial y}{\partial x_2} \delta x_2 \right)^2 + \left(\frac{\partial y}{\partial x_3} \delta x_3 \right)^2 + \cdots \right]^{1/2} \tag{5.23}$$

It is not possible to show the derivation of (5.23).* For most purposes equation (5.22) represents a simple method of calculating errors with a safety allowance.

*See S. J. Kline, and F. A. McClintock, "Describing Uncertainties in Single Sample Experiments," *Mechanical Engineering,* 75, 1953, p. 3.

Example 1

$F = ma$, where

$$m = 2.0 \pm 0.1 \text{ lb}_m \text{ at 5\% significance level}$$
$$a = 30 \pm 0.2 \text{ ft/sec}^2 \text{ at 5\% significance level}$$

Calculate the range of F.
    Nominally,

$$F = 6.0 \text{ lb}_m\text{-ft/sec}^2$$
$$\delta F = \left| \frac{\partial F}{\partial m} \, \delta m \right| + \left| \frac{\partial F}{\partial a} \, \delta a \right|$$
$$= a \, \delta m + m \, \delta a$$
$$= 3.0 \times 0.1 + 2.0 \times 0.2 = 0.7$$
$$\therefore F = 6.0 \pm 0.7 \text{ lb}_m\text{-ft/sec}^2$$
$$\text{Percentage error} = \frac{\delta F}{F} = \frac{0.7}{6} = 11.6\%$$

The value of $\delta F/F$ from equation (5.23) is $0.25/6 = 4.17\%$.

Example 2

Using equation (5.22), if $y = x - z$, then

$$\delta y = \left| \frac{\partial(x - z)}{\partial x} \, \delta x \right| + \left| \frac{\partial(x - z)}{\partial z} \, \delta z \right|$$
$$= \delta x + \delta z$$

The percentage error is

$$\frac{\delta y \times 100\%}{y} = \frac{(\delta x + \delta z) \times 100\%}{x - z} \tag{5.24}$$

Using equation (5.23),

$$\frac{\delta y}{y} = \frac{(\delta x^2 + \delta z^2)^{1/2}}{x - z} \tag{5.25}$$

If the value of x is very close to that of z, the denominator in equations (5.24) and (5.25) can be very small with respect to the numerator and the relative error can become large. This illustrates the fact that care should be taken to avoid if possible the differencing of two nearly equal quantities. It is common to have errors of this sort occur when using computers since the occurrence of this problem cannot be noticed while the machine is performing its various calculations.

Approximate numbers. Accountants and bookkeepers deal with exact numbers. $14.73 means exactly fourteen dollars and seventy-three cents. Most of the numbers engineers encounter are *approximate numbers*. A little reflection on the combination of this adjective and this noun reveals that it is a contradiction in terms. Numbers are always exact. The symbols 14 in mathematical operations mean precisely the same as 14.0 or 14.000000000 or 14.0.

When numbers are referred to as approximate, this qualification is used in either of two meanings which must be sharply distinguished from each other.

1. *Approximate* indicates that the number noted does not differ much from a known true value.
2. The term refers to the process (e.g., measurement) by which the number was obtained.

The results of measurements are always *estimates* of unknown "true" values. Let us first consider case 1 where the true value is known. The only way of recording the value of the ratio circumference/diameter of a circle is by using the symbol π. One cannot give its true value with a finite number of digits. Both 3.14 and 3.141, 592, 653, 589, 793, 238, 462, 643 are approximate values for π.

There are two ways of expressing approximations to true values of the desired degree of precision. The first method, which is used by computers, is *truncation*, i.e., chopping off at the desired decimal place. One can write, for example, $\pi = 3.141$. This means that π lies with certainty within the range 3.141–3.142. Some writers refer to this notation as a *range number*.

On making calculations with range numbers the boundary values should be considered as exact numbers. Otherwise deliberate arithmetical mistakes are added to an already approximate value of π. Therefore, π^2 lies in the range 9.865881–9.872164, or with the sacrifice of little information $9.865 < \pi^2 < 9.873$.

The second method of expressing approximations to true values is by *rounding off.* In this case $\pi = 3.142$. It is tacitly assumed that the reader knows this to mean $\pi = 3.142 \pm 0.0005$. This notation is unfortunately indistinguishable from that for a truncated number. It is generally understood that if a number is not exact, it is arrived at by rounding off the desired decimal place and it is referred to as a *significant number.* The maximum inaccuracy in a significant number is $\pm\frac{1}{2}$ in the last recorded digit.

We can carry out mathematical operations with significant numbers as if they were range numbers. For example, $3.1415 < \pi < 3.1425$ and hence $9.8690225 < \pi^2 < 9.87216425 \pm 0.003142$, which is, of course, equivalent to the range number for π^2.

It is not generally realized that computations with significant numbers lead to results which are *not* themselves significant numbers. The two-digit significant number approximation of π is 3.1. This is understood to mean that π lies in the range 3.05–3.15. Hence, π^2 is in the range 9.3025–9.9225. This range *cannot* be written as a single significant number.

It is common practice to compute with significant numbers as if they were exact and to round off the answer so as to retain the same number of digits as the original number. The result is then considered to be a significant number. This is *wrong*! Following this procedure in our example one obtains

$$\pi^2 = 3.1^2 = 9.61 = 9.6, \text{ i.e., } 9.6 \pm 0.05; \text{ actually, } \pi^2 = 9.8696 \cdots$$

That an erroneous answer for π^2 is obtained is not the result of having chosen a significant number for π only two digits. The significant numbers 3.14 and 3.142 also yield wrong answers for π^2, namely 9.86 ± 0.05 and 9.872 ± 0.0005, respectively. This practice of handling significant numbers need not lead to false answers, but it may do so.

It is, therefore, worth repeating that

1. Arithmetic operations with significant numbers do not yield answers which themselves are significant numbers.
2. The practice of performing calculations with a significant number as if it were an exact number and rounding off the answers so that the same number of digits is retained as in the original significant number may lead to erroneous results.

The results of measurements are never exact numbers. They are estimates of an unknown "true" value. Hence, they are neither range numbers nor significant numbers because it is not certain that the "true" value lies within the measured range. Nevertheless, results of measurements are most often recorded as a single number, and it is tacitly assumed that this is a significant number.

Two remarks are in order:

1. Statements 1 and 2 made above with regard to significant numbers are also valid for the quasi-significant numbers which record the results of measurements.

2. The closeness of approach of the estimate resulting from measurements to the "true" value depends on many factors. Recording a single number for the estimate seems to imply that

> The possible error a trained observer may make will not exceed ½ the unit which corresponds to the last significant digit in the measure. That this is an unwarranted degree of optimism can be readily illustrated by an example. Crookes made ten determinations of the atomic weight of thallium and gave his results (in 1874) as 203.628, 203.632, 203.636, 203.638, 203.642, 203.644, 203.649, 203.650, 203.666. Now Crookes was certainly a trained observer, and if we discard one decimal place, all his values lie between 203.63 and 203.67; yet if we consult a modern list of atomic weights, we find that of thallium

given as 204.39. The history of science is full of other illustrations of the fact that true values may differ greatly from the estimates of them based on measurements, even when those measurements have been made by trained observers.[*]

It is most desirable that the results of measurements be given together with an estimate of the precision of these results. If a single measurement is made to obtain the result, only a subjective estimate of the precision can be made. Of course, there should be some relation between the number of decimals of the numerical result and the estimated precision. One may measure the length of a cigarette with the best caliper available. However, in view of the physical properties of cigarettes, it is not recommended that this length be reported as 101.25 ± 1 mm even though this result strictly speaking contains more information than 101.2 ± 1 or 101 ± 1 mm.

When *many* results of repeated measurements are available, it is appropriate to apply the theory of errors for calculating a range for the numerical result as well as for selecting means to characterize the degree of confidence applicable to this range. This theory and its application was dealt with in the first part of this section. The selection of the type of range and the index of precision must be suited to the type of data and the number of measurements. It is advisable that the engineering student familiarize himself with the Recommendations for Presentation of Data and of the Limits of Uncertainty of an Observer Average published by the American Society for Testing and Materials.[†]

5.8 Interpolation and Extrapolation

Interpolation is the use of data *between* known points, and extrapolation is the obtaining of data *outside* known points or regions. Both must be done (if at all) with considerable care. This is especially true of extrapolation.

The lines obtained by the least squares method are one way of obtaining a suitable complete line on which to interpolate readings. A more sophisticated type of interpolation is shown in Figures 5.18 and 5.19, which show diagrams used in calculating pressure losses in rough commercial piping with fluids flowing through the pipe (the meanings of the terms are not important for this illustration). The first diagram shows the experimental data upon which the second diagram is based. Note in Figure 5.18 that at a value of R of 10^6, six values of ϵ/D have been measured. In Figure 5.19 in the same range, 21 values of ϵ/D have been plotted. The additional values have been interpolated. The dangers of interpolation are illustrated in Figure 5.18 also. In the range from R = 2×10^3 to 3×10^3

---

[*]R. G. Stanton, *Numerical Methods for Science and Engineering,* Prentice-Hall, Inc., Englewood Cliffs, N. J., p. 4. Quoted with permission.

[†]"A.S.T.M. Manual on Quality Control of Materials," Parts 1 and 2, *Special Technical Publication 15-C,* American Society for Testing and Materials, Philadelphia, 1961.

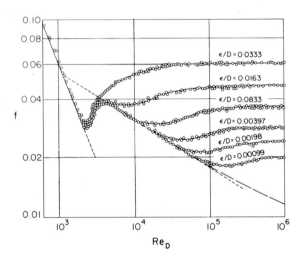

Fig. 5.18. Experimental data for pipe flow losses.

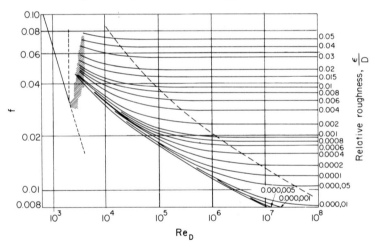

Fig. 5.19. Interpolated data for pipe flow losses.

it would obviously be most dangerous to assume values without measurement. Notice that the investigator has made a large number of measurements at this point. Another example is given in Figure 5.20, which shows the vertical (lift) and horizontal (drag) forces on an aircraft wing. An angle of attack of 22° obviously represents a particular problem, and if data were missing in the range of 16–24°, it would be most dangerous to make assumptions.

The field of fluid flow is a particularly fruitful source of irregular curves. Figure 5.21 illustrates the magnitude of forces (drag) acting on a circular cylinder.

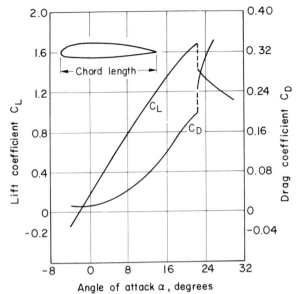

Fig. 5.20. Typical lift and drag coefficients for an airfoil.

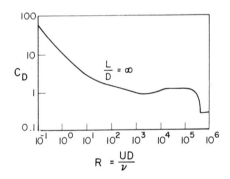

Fig. 5.21. Drag coefficients for circular cylinders.

Notice that if data were available up to an R of 10^3, it would be natural to assume that for larger values of R the curve leveled off to a constant value. Also the "kink" around R = 30 could probably be "reasoned away." The dangers of extrapolation are obvious in cases such as these. A particular case where extrapolation is a great temptation is on time curves. The desire to predict the future is obvious. Two examples should illustrate the point:

1. If the growth of Canada and the growth of the city of Toronto are compared, then by the year 2000, 110% of the Canadian population will live in Toronto.

2. In Section 2.6 it was pointed out that by the year 2200 the mass of scientific publications will be equal to the mass of the earth at present growth rates.

5.9 Exercises

5.1. For the following data, perform the regression of y upon x:

x	y
1.0	0.501
1.5	0.74
2.0	1.07
2.5	1.27
3.0	1.40
3.5	1.78
4.0	1.94

5.2. Given the data

x	1.0	2.0	3.0	4.0	5.0
y	1.1	1.9	2.9	3.9	5.05

calculate the best fit line for the data by regression of y upon x.

5.3. Calculate the equation of the least squares line and the correlation coefficient for the following sets of values: $(x, y) = (3, 2), (0, 1), (0, 0), (-1, -1), (-3, -3)$.

5.4. To find an equation relating the height of a stack of paper and the number of sheets in the stack, the following data were obtained from measurements:

Number of Sheets in Stack of Paper	Height of Stack (mm)
103	9.6
120	11.2
146	13.6
184	17.1
215	20.0

Determine the slope of the straight line that best fits a plot of height of a stack of paper (y axis) against the number of sheets per stack (x axis).

5.5. Group the following variables into dimensionless groups:

    (a) L (length in feet),
      g (gravitational acceleration in ft/sec^2),
      t (time in seconds).
    (b) ω (angular velocity radians/sec),
      V (linear velocity ft/sec),
      R (length in ft).
    (c) V (velocity in feet/second),
      L (length in feet),
      g (gravitational acceleration ft/sec^2).
    (d) P (pressure in pounds force/ft^2)
      V (velocity in feet/sec)
      ρ (density in lb force)(sec^2/ft^4).

5.6. Propose at least three ways of plotting the variation of temperature with height over the first 10 ft above the ground.

5.7. When in the process of measurement should numerical values be rounded off? Round off the following numbers to three significant figures: 27.65, 27.651, 27.55, 27.551, 27.499, 2765, 27,651, 27.550, 0.027,551, 0.002,749,90.

5.8. For $x = 5 \pm 0.1$, calculate the relative errors for z in the following cases:
    (a) $z = x^3$.
    (b) $z = x^2 + 2x + 1$.

5.9. For

$$a = 1.0 \pm 0.1$$
$$b = 2.0 \pm 0.1$$
$$c = 3.0 \pm 0.1$$

calculate the relative errors in z for the following:
    (a) $z = abc$.
    (b) $z = a^2 + b^2 + c^2$.

5.10. If $z = x^2 + 3x + 2$ and we require z to be known to 1% precision in the range of x near 2, to what precision must we know x (in percent)?

5.11. The construction error in a volt-ammeter is specified by the manufacturer as follows:

    1. Voltage: 2% of indication between full-scale and half-scale deflection; below half-scale deflection, 1% of full-scale value.
    2. Current: 1% of full-scale value of the effective range.

    (a) For the 100-v range, sketch a graph of the absolute error ΔV v versus voltage and a graph of the percentage error $\Delta V/V$ versus voltage.

(b) For the 100-ma range, sketch a graph of the absolute error ΔI versus current and a graph of the percentage error $\Delta I/I$ versus current.

What conclusions may be drawn from the results of parts (a) and (b) with regard to measurement practice?

5.12. A meter stick is subdivided into millimeters:

(a) What is the reading error?
(b) If an object is measured and found to be 700 mm, give the percentage error arising from the reading error.

5.13. Prove that the right-hand sides of equations (5.13), (5.14), and (5.15) are equal.

5.10 Suggestions for Further Reading

Bridgman, P. W., *Dimensional Analysis*, Yale University Press, New Haven, Conn., 1931.

Kline, S. J., and McClintock, F. A., "Describing Uncertainties in Single Sample Experiments," *Mechanical Engineering,* 75, Jan. 1953.

Langhaar, H. L., *Dimensional Analysis and Theory of Models,* John Wiley & Sons, Inc., New York, 1951.

Schenck, H., Jr., *Theories of Engineering Experimentation*, McGraw-Hill Book Company, New York, 1961.

Wildi, T., *Units*, Volta Inc., Quebec City, P. Q., Canada, 1971.

6

Reporting Results

Original work is not complete until it has been reported. This reporting must be done as clearly and concisely as possible. In addition, in technical and scientific work it is necessary not only to make your conclusions clear to the reader but to make them inescapable. In other words, you are required to prove them by presenting your evidence.

6.1 The Technical Report

The standard method of reporting scientific or technical results of any sort is the technical report. Examples are internal company reports and memoranda, scientific and technical papers, and technical and semitechnical magazine articles. The general format of these presentations is quite restricted and almost all conform to it, although the order may be changed and certain sections left out depending on the content. The sections of a technical report have the titles of the sections of this chapter (Sections 6.2-6.11), and each will be discussed separately. This restriction on format does not imply that an author does not have considerable freedom in his writing method. For example, it is conventional practice (but only in the last 30 years) to write technical reports exclusively in the third person (to use "the authors" other than "we" or "I"). There is no necessity for this. However, writing technical reports in the first person without sounding either pompous or sloppy is quite difficult. A budding author is advised to confine himself to the third person, at least in the beginning.

The standards of technical reports are the same as for any other form of writing. In particular, brevity and clarity are important. The importance of the various sections of the report depend on the particular way in which they are read. In this chapter we shall not deal with style in writing since this is the subject of many useful books. A selection of them is listed at the end of this chapter.

As laid out within this chapter several sections of the report are optional. Only the abstract is required as a separate entity. The various sections may quite easily exist simply as separate paragraphs of a short report. The order is not necessarily fixed, except for the conclusions.

Readers of reports can be roughly classified according to the nature and the degree of their interest. Colleagues, collaborators, and others working in your field will have a technical interest in what you have written. In industry a good immediate supervisor will have an interest in your work, from a technical point of view as well as for managerial reasons. If your reports are read at higher levels of the organizational hierarchy, the motives for reading will mainly be managerial rather than technical.

What fraction of your report is actually read depends foremost on the time available to the reader. Any professional man must read some of the new technical literature. If he does not, his knowledge will be obsolete in a short time. Keeping up with new developments is an increasingly demanding task. This is illustrated in Figure 6.1. Since World War II, the number of abstracts published by one professional society increased at the rate of 9%. Therefore, reading time is becoming increasingly valuable. Report reading competes with other time-consuming activities of your colleagues and superiors. If you want your reports to be read, they must be of high quality.

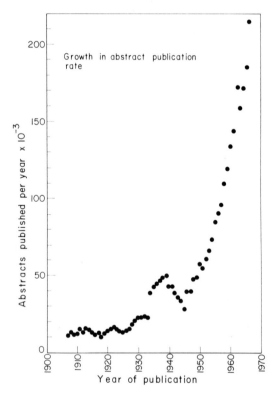

Fig. 6.1. Growth in abstract publication rate.

Which parts are read, and the intensity of the reading, perusal, or scanning, is mainly determined by the nature and the degree of the reader's interest. In Table 6.1 an attempt is made to illustrate this point.

Table 6.1.   Sections of Report Read as a Function of the Nature and the Degree of Interest to the Reader

Degree of Interest	Nature of Interest		
	Technical	Supervisional	Managerial
High	A W F T C	A W F T C	A C
Medium	A F T C	A C	A
Low	A	A	–

Key: A = abstract; W = written text; F = figures, graphs; T = tables; and C = conclusions.

The assessment condensed in Table 6.1 is, of course, only qualitative. However, it should be clear that one of the most important parts of your report, if not the most important section, is the summary. There are far more people with a marginal interest in the subject of your paper than there are fellow professionals for whom your findings are of direct, immediate interest. Therefore, the summary, abstract, or synopsis is the most frequently read part of your report.

As a result of the enormous increase in the volume of technical publications, many organizations have set up abstracting services for their members or on a commercial basis. Often the summaries of papers are collected and sorted according to the technical area of interest and distributed or sold. This growing organized approach to information retrieval and dissemination justifies expanding extra effort in choosing the proper title of reports and papers and in writing a summary which is at the same time condensed and comprehensive. The abstract should be expressed in lucid terms. It must be as brief as is consistent with clarity. There are many books on the subject of clear writing in your library. Many of these books refer to particular kinds of writing, such as term papers, short stories, and technical reports. The principles of good writing are universal and the details appropriate to special types are easily learned.

6.2 Front Matter

The front matter includes the title, title page, letter of submittal, table of contents, and nomenclature. We shall deal separately with each of these items.

"Friction" is too short a title. "The Resistance of Seventeen Ball Bearing Races to Rotation in the Range 5 to 5000 Pounds of Loading and 10 to 2000 Revolutions Per Minute" is too long. Since a large number of casual readers of reports will decide whether or not to read further, the title should be as informative as possible. "Resistance to Rotation of Ball Races Under Load" gives the essential information. Titles of technical reports are typically longer than in other types of written material. The title page includes, in addition to the title, the authors and their position, the date, and such information as for whom the work was done, the publisher (if printed), and the copyright information if copyrighted.

The letter of transmittal is the letter or memorandum which accompanies any report or article stating in effect that the job is done and referring back to the original request. If sent to a superior, it will serve to remind him of the date and circumstances of his request to you.

The table of contents outlines in order the topics covered in the report, in the order of the report. Tables of contents are occasionally omitted for short reports and articles. The method of arranging a table of contents is identical to that in a book.

If algebraic symbols are required in the report, the terms should be defined when they are first used. If any large quantity of these terms is used, they should also be defined in a separate nomenclature list. In this list the terms should be listed in order alphabetically for the Roman characters and then Greek and German characters. Subscripts and superscripts may be listed separately.

Here is an example of nomenclature used in a technical paper on the subject of extrusion:

Notation

$A(\alpha)$	defined by equation (3a)
$B(\alpha)$	defined by equation (3b)
H	lubricant film thickness
N	modified Sommerfeld number, $(\eta V_1)/(\bar{\sigma} r_1)$
P_1	axial extrusion load
p	lubricant pressure
$\bar{p}$	dimensionless extrusion pressure, $p_1 (\pi r_1^2 \bar{\sigma})$
r	radius
r_1	initial radius of extrusion at entry
r_2	final radius of extrusion at exit
s	coordinate measured along die taper
V	slip velocity of extrusion relative to die
V_1	axial velocity of extrusion at entry to die
W_p	rate of plastic work
W_s	rate of shear work in lubricant
α	semicone angle of die
η	lubricant viscosity
λ	pressure coefficient of viscosity defined by $\eta/\eta o = \exp(\lambda p)$
$\bar{\sigma}$	yield stress
τ	shear stress

6.3 The Abstract

Unlike a novel or short story, a technical report has no climax. The reader should be told of the basic accomplishment and the conclusion as soon as possible. This is done in the abstract. An abstract or summary should seldom be longer than one typed page and typically is a single short paragraph. *Scientific American* magazine confines it abstracts to three or four lines. Here are some titles and abstracts from one issue of the magazine.*

1. *Lead Poisoning*
   Among the natural substances that man concentrates in his immediate environment, lead is one of the most ubiquitous. A principal cause for concern is the effect on children who live in decaying buildings.

2. *The Iroquois Confederacy.*
   This alliance of Woodland Indian tribes played a significant role during the European colonization of North America. Excavations in New York now cast new light on their origins and social evolution.

3. *The Fast Computer.*
   ILLIAC IV is made up of 64 independent processing units that by operating simultaneously will be capable of solving complex problems in a fraction of the time needed by any other machine.

4. *The State of Water in Red Cells.*
   Water in these cells has long seemed to exhibit anomalous properties. Recent studies show that it is the same as other water; the apparent anomaly arises from the close association of hemoglobin molecules.

Here are examples of longer abstracts taken from two technical journals:

1. W. H. Melbourne, *Wind Tunnel Tests on a Radio Tower,* Australian Defence Scientific Service, Department of Supply, Aeronautical Research Laboratory, Melbourne, Australia, Aerodynamics Note 251, 8 pp. + figs (1965).†

   The design of a fabricated 250-ft.-high radio tower by the State Electricity Commission of Victoria incorporates a lift well and stairs enclosed in a square center section. It was thought that the tower cross section was such that a galloping type of oscillation could possibly be excited.

   Measurements were made in a wind tunnel of the lift and drag of a section of the tower. Using a quasi-steady analysis, it was shown that the galloping type of oscillation would be excited and that the estimated structural damping of the tower would not be sufficient to safely contain the oscillations to small amplitudes. Further wind tunnel tests indicated that the addition of flat plate spoilers along the edges of the square center section reduced the aero-

---

Scientific American, 224, No. 2, Feb. 1971. Quoted with permission.
†Quoted with permission.

dynamic excitation to a level which could be safely absorbed by the estimated structural damping for wind speeds up to 100 mph.

2. R. Bell and M. Burdekin, *Dynamic Behaviour of Plain Slideways,* Institution of Mechanical Engineers, London, Applied Mechanics Group P8/67, 13 pp. (1966–1967).*

The friction characteristics resulting from the motion of one surface over another form a very important facet of the behaviour of many physical systems. This statement is particularly valid when considering the behavior of machine tool slideways. Most slideway elements consist of two plain surfaces whose friction characteristic is modified by the addition of a lubricant. In many cases, the complete slideway consists of many mating surfaces and the choice of slideway material, slideway machining, and lubricant is often influenced by the long term problem of wear. The aim of this paper is to present results of experiments on a test rig designed to be representative of machine tool slideway conditions; the experiments were wholly concerned with the behaviour of the bearing under dynamic conditions. The major emphasis is on results obtained with a polar additive lubricant which appears to exclude the possibility of "stick-slip" oscillations. A parallel series of tests are reported where a normal hydraulic oil was used as lubricant. The use of this second lubricant allowed some study of the "stick-slip" process. The dynamic friction characteristics cyclic friction characteristics and damping capacity of several slideway surface combinations have been obtained and are discussed in the context of earlier work in the field and the role of slideways in machine tool behaviour.

Several points may be made from these articles:

1. The abstract stands by itself. References to previous work are not included, and so no references are required at the bottom of the abstract.

2. For the same reason, mathematics is seldom included, since this would require the defining of every symbol. Occasionally, as in a purely mathematical paper, mathematical terminology is unavoidable.

3. Considerable care is taken to exactly define what was done. In the first abstract theory was required; the solution was "flat plate spoilers," the tower was 250 ft high, etc. In the second example only two types of oil were used. The tests were done on an experimental apparatus and not on an operating machine; only specific named phenomena were considered.

6.4 The Introduction

This is the section in which to place the work in perspective. The reason for doing the work (the statement of objectives) should be placed here. The work

*Quoted with permission.

of other people can be outlined and related to your own work. This should be done briefly since most references used should be available to the reader if he wants more detail. In this section you can, for example, state that

1. You are attempting to develop an analysis as well as do experiments on the problem.
2. Your particular study takes a new approach or is more accurate than previous work.
3. This is an attempt to check out someone's analysis.
4. This is a new piece of equipment being tested.
5. This is an attempt to produce new design data.

The introduction is seldom a historical essay. It should typically be addressed to a level of knowledge similar to your own when you began the project. This implies that it should not include references proving, for example, that mass is conserved, nor should it assume that the reader is familiar with the details of all of your references.

6.5 Presentation of Theory

If it is necessary to have mathematical analysis in the report, a separate section is frequently devoted to it. It can also be part of the introduction or simply in an unlabeled section of the report. It is useful to review briefly an existing analysis if it is important to your work. If the analysis is your own or a variation on an existing one, then it should be given in sufficient detail that a reader can repeat it himself. All symbols must be carefully defined where they are first used. This is usually done in the separate nomenclature section as well as in the body of the report.

6.6 Presentation of Experiments and Results

In this section we are referring only to the written part of the report. Most results are presented in graphical, tabular, analytic, or similar detailed form. This is the subject of Section 6.11.

In the written part of the report you should make as much use of diagrams, graphs, photographs, etc., as possible in order to shorten the written descriptions—"One picture is worth. . . ." Topics which you should cover in this section include

1. Description of equipment.
2. Ranges of variables in the tests.
3. Any fundamental difficulties encountered, although no one wants to hear about your day-to-day problems.
4. Your estimate of the accuracy and precision of your results.
5. Your method of reducing the raw data to their final form.
6. Explanations of the data; point out anomalies, regions of increased error, areas where new variables come into play, etc.

6.7 Comparisons Between Theory and Experiments

This section is used to compare any analytical models used and the experimental results. It is the place to attempt to explain discrepancies between them upon the basis of errors, oversimplification, scatter of results, incorrect analysis, unmeasured variables, etc. In this section, empirical curves are put through experimental data, and the analytical part of the work is justified or rejected.

6.8 Conclusions

This is the place to refer back to the original objectives. The conclusions should be implied by the work in the body of the report and should therefore be brief.

Examples of Conclusions

1. J. Lenihan, "Turning Rubbish into Alcohol," *Engineering Digest,* Aug. 1970.*

   Increasing prosperity puts more paper and vegetable waste (also a good source of cellulose) into the dustbins. Now, for the first time, there is a real possibility that engineers and chemists cannot only eliminate the growing cost of refuse disposal but can actually make this unpleasant process yield a worthwhile profit.

2. J. S. McCabe and E. W. Rothrock, "Multilayer Vessels for High Pressure," *Mechanical Engineering*, March 1971.*

   Multilayer vessels have extended the art of pressure-vessel construction and presented the process designers with equipment useful in a wide range of operating conditions.

3. R. F. Jurgens, "Radar Scattering from Venus at Large Angles of Incidence and the Question of Polar Ice Caps," *Science*, Dec. 20, 1968.*

   The results of this study and the previous study by Jurgens clearly indicate homogeneous scattering properties of Venus over a range of incidence at least as large as $75°$ and that no latitude dependence is observable. Therefore, ice caps do not exist on Venus unless the very unlikely situation has occurred and their radar-scattering properties at large incidence angles are not significantly different from those of the equatorial region. Large fields of broken ice slabs could duplicate the radar-scattering behaviour of the equatorial region at large angles of incidence if the surface roughness were just slightly greater than the roughness of the equatorial material, but snow is excluded by these radar observations.

*Quoted with permission.

4. C. Kornetsky and G. Bain, "Morphine: Single Dose Tolerance," *Science*, Nov. 29, 1968.*

These results confirm the phenomenon of single-dose tolerance in the rate and indicate that, at these doses, the tolerance is not presented [until at least] 24 hours after the initial dose was given. In fact, the tolerance becomes more pronounced the longer the time interval between the two doses of the drug. The mechanism for this type of tolerance may be quite different from that after repeated large doses of morphine sulphate.

One hypothesis that has been preferred is that this single dose tolerance may be the result of an immune mechanism. However, these experiments have not given consistent results. In fact, there is some evidence suggesting that there may be a potentiating factor in the serum of the animals previously made tolerant to morphine sulphate. Thus, the enigma of tolerance to this drug is not solved but only made more complicated by the results of this experiment.

Example 4 is an excellent illustration of the fact that results are not always what is expected and can be negative insofar as solving the previously defined problem is concerned. Of course, this is seldom done on purpose.

6.9 Acknowledgments

It is common practice in closing a report to list briefly any financial support or facilities made available to the authors. This is the place to acknowledge help, encouragement, guidance, etc., which the author has received. It is not necessary to add a separate section to the report since usually one or two sentences is sufficient.

6.10 References and Sources

The obtaining of reference and resource material has been dealt with in Section 2.6. *All* true reference material should be given in the technical report. Professional and scientific etiquette requires that you do not claim as your own the work of others.

There is a distinction between references and bibliographies. References are actually referred to in the body of the report. Bibliographies are seldom used in a report but are common in books. These refer to a list of writings on a given subject but which are not referred to in the body of the report. In textbooks, the bibliographies are often listed under "Further Reading" or "Additional Material."

*Quoted with permission.

The references may be placed at the bottom of a page of a report, at the end of a section (usually used in books), or at the end of the report. The most common position is at the end of the report.

Several ways of referencing are given in the following sentences taken from scientific journals. Most companies and journals require consistent use of a single method in their material.

Examples

1. "It has been found (3, 4) that. . . ."
   The numbers 3 and 4 refer to references 3 and 4 at the back of the report.
2. "It has been found by Smith (2) that. . . ."
   Smith is the author of reference (2).
3. "This contradicts the conclusions of Smith (1969) and Jones (1958) that. . . ."
   The references will be listed alphabetically at the end of the text by author and year of publication.

The references may also be listed in several ways.

Examples

1. Addison, Herbert: *Hydraulic Measurements*, John Wiley & Sons, Inc., New York, 1946.

This is a book. The information is given in the following order: author, title, publisher, place of publication, date of publication. Edition numbers and page references are often added to this.

2. M. H. Green, *Air Pollution Control Association Journal*, Vol. 16, No. 11, p. 703, December 1966.

Here the information is author, journal title, volume number, issue number, page number, and date. *This type of reference is not recommended.* For brevity, the reference lacks the title of the article. This often makes it difficult for the reader to decide whether or not to refer to the original reference in more detail.

3. Avitzur, B., "Analysis of Metal Extrusion," *J. Eng. for Ind.,* Trans. ASME, Ser. B, *87,* 1, 57–70, Feb. 1965.

Many references use abbreviations, which, if they are not familiar, are often difficult to decipher. These abbreviations, however, are standardized, and only certain forms may be used. Reference sections of libraries keep dictionaries of permitted abbreviations and acronyms. An acronym is a word made of the first letter or letters of a series of words, e.g.:

Radar: Radio detecting and ranging.
Project Wombat: Waste of money, brains and time.

The particular reference given lists author, title, journal (*Journal of Engineering for Industry,* which is Series B of the *American Society of Mechanical Engineers Transactions*), volume number (underlined or italics), issue number, pages of article, and date.

      4. Smith, J. D., Private Communication, 1970.

      It is common to receive important information personally. If you wish to give credit for this (especially if given by letter), then this is the standard format.

      The student should take particular note of the order in which the references are given (author, title, journal, issue, page, date, etc.) and the punctuation. Failure to do this, especially if abbreviations are used, can cause considerable confusion.

6.11 Visual Presentation of Data

      The most common and most efficient ways to present quantitative data are by the use of tables, graphs, mathematical equations, and drawings. Because of the importance of each of these methods, we shall deal with each of them in some detail. Which method of presentation is the most appropriate form of reporting depends on a variety of factors, which include the purpose of the report, the quantity of data, the degree of precision of the results, and the number of variables. Because so many factors are involved, only a few general remarks can be made here. Your choice of the particular form of presentation must be made largely by applying common sense guided by experience.

      The purpose of the report may be technical. In this case, the reader is supposed to be knowledgeable in the field with which the data are concerned. The report should then be comprehensive, precise, and detailed. If, however, it is your intent to inform the readers only summarily or if you address yourself to nontechnical readers, the preferred form of presentation may be more dramatic than complete or precise. In the technical literature of the transportation field, for example, you may find in a paper on the economy of transportation, a list of the various costs of owning and operating an automobile in the form of a table such as shown in Figure 6.2. In a weekly family magazine, however, it would be more appropriate to employ a more pictorial way of representing the same data, e.g., in the form of a pie chart, as shown in Figure 6.3.

      The number of data to be reported will also play a role in your choice of the device of presentation. If a measurement has been repeated many times over, one would rarely record all the results obtained, if only for economic reasons. In such a case it is more expedient to condense the data to some average value and to record this value along with some indication of its precision. When a large number of different types of measurements have been carried out on a large

Distribution of Costs of Owning and Operating a Typical Automobile			
		%	Total %
Direct Costs	Depreciation	28.0	
	Repairs and maintenance	19	
	Gasoline (excluding tax)	14	
	Insurance	13	
	Garage and parking, tolls,	10.0	
	Oil, tires, etc.	4.0	88.0
Indirect Costs	Taxes, Gasoline and other	12.0	12.0
			100.0

Fig. 6.2. Tabular presentation of data.

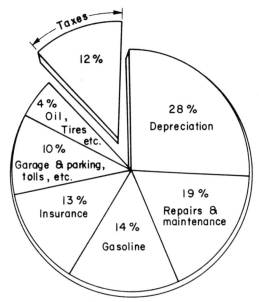

Fig. 6.3. Pie chart showing distribution of costs of automobile ownership.

number of different materials there is no choice but to list all the results. Handbooks contain many tables of this type, as shown in Figure 6.4.

The degree of precision of the results to be recorded is another factor that needs consideration in the choice of the means of reporting. It does, of course, make little sense to plot results obtained with a significant precision of 0.1% on standard-sized (7 × 10 in.) graph paper from which values can be read with a

Name	Density in gm/c.c.
Aluminum	2.70
Antimony	6.618
Barium	3.5
Beryllium	1.85
Bismuth	9.781
Cadmium	8.648
Calcium	1.873
Chromium	7.16
Cobalt	8.8
Copper	8.97

Fig. 6.4. Density of metals at 20°C.

precision of at best about 0.5%. Considerable effort and costs often need to be expended to obtain high precision. So as not to lose this precision on reporting, a table of the results is probably the appropriate form of presentation.

If rather imprecise information has been gathered, it may not be worth spending the time to calculate a measure of precision. A graph of the data will then often suffice to present the numerical values to the reader and to indicate simultaneously the degree of relationship between the variables. Figure 6.5 gives an example. Two types of viscosity are plotted for a group of materials which are chemically similar, differing only in physical structure and molecular weight. The Mooney viscosity is a property of the bulk material; the intrinsic viscosity is measured in solution. The graph shows that two measures of viscosity of a particular polymer are probably related and that the precision of measurement is likely low.

The number of variables which can be displayed sensibly in graphical form is necessarily restricted often to three and sometimes to two. Figure 6.6 demonstrates that three variables in one graph can be too many.

In general, tables are more suited for reporting sets of data with several variables. Tables can accommodate up to four or five variables if the range is not large.

Tables. Corresponding quantitative or qualitative descriptions of dependent and independent variables, orderly arranged, constitute a table. Three different types of tables may be distinguished:

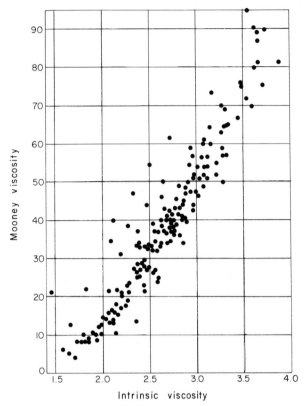

Fig. 6.5. General graphical presentation of data.

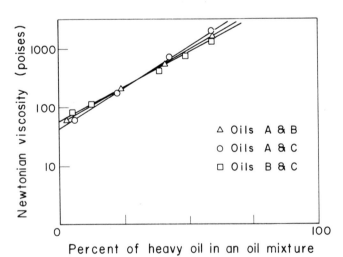

Fig. 6.6. Three variables on a single graph.

Qualitative tables list qualitative information on properties, materials, processes, etc. Figure 6.7 gives an example of such a table.

In statistical tables some variables are listed quantitatively, and others, including the independent variables, are presented qualitatively. Examples are

Table Number.	**Comparison of Bulk and Emulsion Polymerization**	
	Bulk	*Emulsion*
Rate of polymerization	Low	High*
Average molecular weight of polymer	Low	High*
Viscosity of polymerization medium	High	Low*
Content of nonpolymeric materials in product	Low*	High
Recovery process of product	Simple*	Complex
Sensitivity of course of reaction of purity of reagents	Low*	High
*Generally considered to be advantageous.		

Fig. 6.7 Tabular presentation of qualitative data.

Conversion of Pressure Units				
	Dynes/sq cm	*Grams/sq cm*	*Atmosphere*	*lb/sq in.*
Dynes/sq cm	1	1.020×10^{-3}	9.870×10^{-7}	1.450×10^{-5}
Grams/sq cm	9.806×10^2	1	9.678×10^{-4}	1.422×10^{-2}
Atmosphere	1.013×10^6	1.033×10^3	1	1.470×10
lb/sq in.	6.894×10^4	7.031×10	6.805×10^{-2}	1

Fig. 6.8. Table for the conversion of pressure units.

the table of atomic weights, tables of physical constants of various types of materials, recipes, lists of weights and measures, and tables of conversion factors, as shown in Figure 6.8

Some terms pertaining to tables are given in Figure 6.9. Quantitative tables list the data of all variables in numerical form. Usually the variable chosen as the independent one is listed in the stub. An example is shown in Figure 6.10.

In the functional class of tables, relations of the type $y = f(x)$ are tabulated. A functional table is reproduced in Figure 6.11. Tables of logarithms of trigonometric functions and the like are other examples of functional tables.

Generally the number of digits included in numerical data listed in tables reflects the precision inherent in the measurement. However, the number of figures used to report a measurement is principally a question of aesthetics. The confidence interval reflects the reliability of the measurement and not the number of figures included.

Table Number.	Title of Table	
Additional remarks can be added in a headnote.		
Stub Heading (units)	*Box Heading*	
	Column Heading (units)	*Column Heading (units)*
Stub	Item	Item
Stub	Item	Item
Stub	Item	Item
Stub	Item	−*

Footnote: In scientific journals vertical lines are often avoided because of the cost of printing. The columns are then spaced widely.
*Information is lacking.

Fig. 6.9. Terms used to describe the components of a table.

**Barometric Pressure
Observed Three Times Daily**

Period March 1952–April 1955			
Pressure Range (mm Hg)	*Number of Observations*	*Frequency (%)*	*Cumulative Frequency*
720–729	5	0.5	0.5
730–739	42	4.0	4.5
740–749	204	19.4	23.9
750–759	293	27.9	51.8
760–769	359	34.2	86.0
770–779	97	9.2	95.2
780–789	48	4.6	99.8
790–799	2	0.2	100.0
	1050		

Fig. 6.10. Tabular example of exhibiting quantitative data.

Growth of Corn Stalk			
Age (days)	*Height (in.)*	*Number of Leaves*	*Weight of plant (lb)*
3	3	1	0.01
15	8	4	0.30
30	24	8	1.00
60	36	12	3.00
90	54	15	6.00
120	72	18	10.00

Fig. 6.11. Tabular presentation of numerical data.

Tables containing exponents are quite common and are easily misread. For example, $P \times 10^3$ at the head of a table and a value of 3.0 gives $P = 0.0030$, not 3000.

Graphs. Often collections of data are depicted in graphs; i.e., the values are represented by areas or by distances to a base line or a set of coordinate axes. This is an appealing method of displaying information. Graphs have some dramatic impact. They also allow easy comparison of data. Graphs, furthermore, indicate visually the precision and/or reproducibility of the measurements and provide some insight into a possible relation between variables.

Two classes of graphs can be distinguished: qualitative and quantitative graphs. The first type is better suited for nontechnical objectives, and the latter category is the one most often used in scientific and engineering reports. There are no hard and fast rules for choosing the appropriate type of graph. Common sense should be your guide, and clarity of presentation your goal.

Examples of qualitative graphs are given in Figures 6.12, 6.13, and 6.14, which show pictographs, a single bar chart, and multiple bar diagrams, respectively. Pie charts belong to this category (Figure 6.3).

In a quantitative graph the results of measurements or other data are shown as points positioned such that the distances relative to two coordinate axes represent their values. The axes usually intersect at right angles. The *vertical* and *horizontal* reference lines are the ordinate and the abscissa axes, respectively. There is a large variety of types of graph paper available on which grids are printed for easy plotting. These are available with 4, 5, 6, 8, 10, 12, and 16 lines/in. or 5 and 10 lines/cm in grid sizes 7×10 and 10×15 in., 18×25 cm and 25×38 cm, or in rolls.

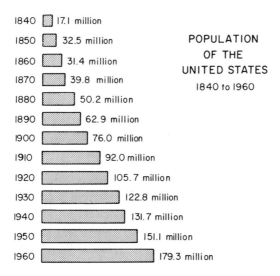

Fig. 6.12. Qualitative graph.

Sources of student funds

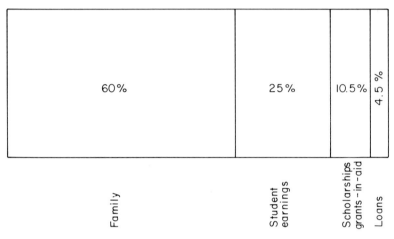

Fig. 6.13. Qualitative graph.

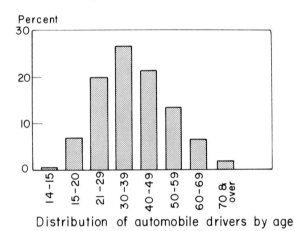

Distribution of automobile drivers by age

Fig. 6.14. Qualitative graph.

An even greater variety of paper with rectangular grids exists on which the distances between lines and/or the number of lines between heavy grid lines in the ordinate direction differ from those in the abscissa direction. This large variety allows the appropriate choice of paper when plotting quantities expressed in units that have nondecimal subunits: gallons, quarts, yards, feet, years, months, etc.

Circular grid paper is useful for the plotting of trigonometric data or results expressed in polar coordinates (distance and angle). This type of paper is shown in Figure 6.15. The intensity of light is proportional to the distance from the origin (+) and is plotted as a function of direction.

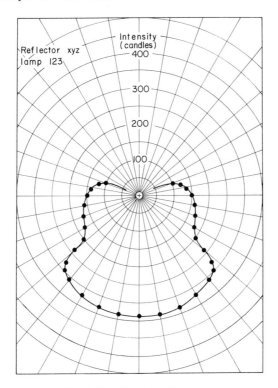

Fig. 6.15. Circular grid paper.

On a triangular grid, data may be represented which depend on three variables, the sum of which is constant, e.g., compositions of a mixture of three components. An example is given in Figure 6.16. This graph shows which mixtures of phenol, acetone, and water are completely miscible, i.e., form one phase, and which are not miscible. Each point in this type of diagram represents a particular composition of the mixture. The relative amounts of phenol, acetone, and water represented by point P are measured by the distances PA, PB, and PC, respectively, and are read on three scales at the intersection of the appropriate gridlines Pa, Pb, and Pc. The composition corresponding to P is 25% phenol, 15% acetone, and 60% water. Since P falls in the area encompassed by the curve and the base line, the mixture indicated by P is not miscible.

A third type of paper, namely functional graph paper, will be described on page 124.

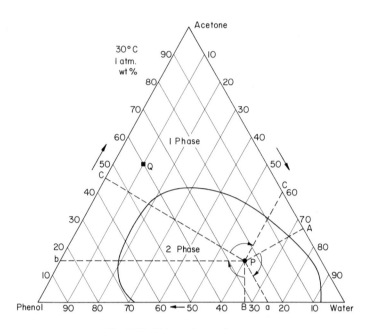

Fig. 6.16. Triangular graph paper.

The following remarks may be of assistance in the preparation of graphs:

1. Choose paper of a size and with distances between grid lines in accordance with the precision of the data.

2. It is customary to measure the independent variable along the abscissa.

3. Choose the coordinate scales and their value ranges such that the data points cover as large an area of the paper as possible.

4. It is recommended to assign values to successive heavy grid lines which differ by 1, 2, 4, or 5 units.

5. If the precision of the data is about the same in the ordinate and abscissa directions or is irrelevant or not known, the units and the range of the scales should be chosen such that the data points or curves cover a near-square area. If, on the other hand, the precision of the data in the ordinate direction differs appreciably from that in the abscissa direction, the ratio of height to width of the area to be covered by the data points should as much as practicable reflect the ratio of the precisions in the ordinate and abscissa directions.

6. Graphs are not complete without a fully descriptive caption, the name of the quantities represented along each axis, and the units in which these quantities are measured. Furthermore, the main grid lines must be marked with the values assigned along the coordinate axes. For the sake of clarity it is to be recom-

mended that these numbers labeling the divisions along the axes have a small number of digits. Do not label an axis as follows:

$$0.0020 \quad 0.0025 \quad 0.0030 \quad 0.0035$$
$$\text{Temperature Difference } (^\circ\text{C})$$

Instead write $2.0 \quad 2.5 \quad 3.0 \quad 3.5$

$$\text{Temperature Difference} \times 10^3 \; (^\circ\text{C})$$

or

$$\text{Temperature Difference} \; (^\circ\text{C}) \times 10^3$$

(The notation "Temperature Difference, $^\circ\text{C} \times 10^3$ is somewhat ambiguous and should be avoided.) When the logarithm of a quantity is plotted, the units of this quantity must be indicated, although logarithms themselves are dimensionless numbers, e.g.,

$$0 \quad 1 \quad 2 \quad 3$$
$$\text{Log Time (hr)}$$
$$10^0 \quad 10^1 \quad 10^2 \quad 10^3$$
$$\text{Time (hr)}$$

Both these forms of notation are in use.

7. Sometimes experimental data are compared in graphs with computed curves. To clearly distinguish between data and curves it is good practice to omit the computed points through which the curves necessarily pass.

Equations. If two quantities are related, the most compact and convenient form of expressing their relation quantitatively is that of a mathematical equation. The existence of a relation known from past experience or expected from theoretical considerations is indicated by the pattern of data points in a graph. If we assume that a linear relation of the form $y = a + bx$ is indicated between x and y, where a and b are constants, we can deal with the problem by the method of least squares described in Chapter 5.

If a graph of two quantities indicates that there is a nonlinear relationship between the two, one is faced with the problem of selecting a nonlinear equation $[y = f(x)]$ to represent the relationship. When neither experience nor theory offers guidelines, a process of trial and error must be followed to choose the appropriate form of $y = f(x)$. In this process, functional graph paper is helpful.

On this type of paper the grid lines in one or both directions are spaced over distances proportional to the values of a given function $y = f(x)$ for conventionally chosen series of values of x. Certain simple mathematical forms of the function y occur, or are presumed to apply, sufficiently frequently to merit special attention.

1. Exponential Relation

Many theories in physics lead to exponential relations between variables. The general mathematical expression for these relations is

$$y = a \exp(bx) = ae^{bx}$$

where a and b are constants. Then $ln\, y = ln\, a + bx$ or $\log y = \log a + b'x$ with $b' = 0.4343b$. Therefore, a plot of $\log y$ (ordinate) versus x (abscissa) should yield a straight line with an intercept of the ordinate axis of $\log a$ and with slope b'. For this case, semilogarithmic graph paper has been designed. Along one of the axes the distances of the intersections with the grid lines to the origin are proportional to the logarithm of the numbers marked along this axis. The other axis has arithmetic (i.e., equidistant) divisions. Thus a plot of y versus x can be made directly on this type of paper without first finding the values of $\log y$ in tables. An example is shown in Figure 6.17, which demonstrates that the population of Canada increases exponentially. One may expect with fair confidence that in 1980 the population will be 25 million. This number can be read directly from the graph.

Depending on the range of values of y to be plotted, one or more logarithmic cycles are required. This number of cycles (n) is equal to the nearest higher integer value of $\log (y_h/y_l)$ where y_h and y_l are the highest and lowest value of y, respectively. Graph papers with one to seven logarithmic cycles are available.

2. Power Law Relation

Some theories lead to a relation of the type

$$y = ax^b \quad \text{(power law)}$$

Many empirical relations are also expressed in this form. Here $\log y = \log a + b \log x$; hence, a plot of $\log y$ versus $\log x$ should yield a straight line. For convenient plotting of data related by a power law log-log graph paper can be used on which both axes are marked with divisions in logarithmic proportions. A variety of graph papers with grid patterns from 1×1 to 3×5 logarithmic

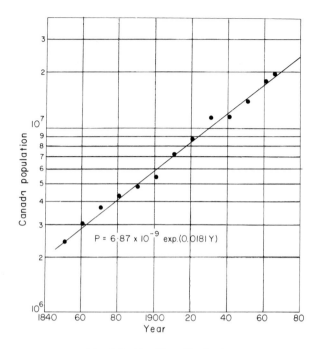

Fig. 6.17. Semi-logarithmic plot.

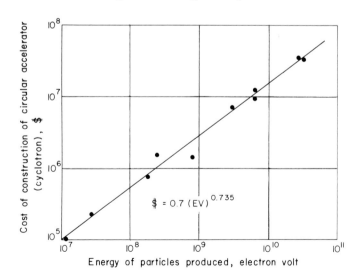

Fig. 6.18. Log-log plot.

cycles is available. Figure 6.18 shows an example of a log-log plot relating the cost of building cyclotrons to the energy of the particles to be generated.

3. *Probability Function* or the *Normal Curve*

It is often desired to test whether a series of results from repeated experiments follows the law of normal frequency distribution. Let us consider the data reproduced in Figure 6.10. The relative frequency of occurrence of barometric pressures in the 10 mm Hg ranges are shown in Figure 6.19 in the form of a histogram. From the data, the mean value and the standard deviation were calculated: $\bar{x} = 758.8$ and $\sigma = 11.6$ mm Hg. The normal frequency curve corresponding to these numbers is

$$F(x) = \frac{10}{11.6\sqrt{2\pi}} \exp \frac{-(x - 758.8)^2}{2 \times 11.6^2}$$

where x is the barometric pressure.

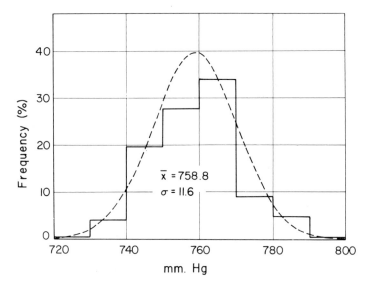

Fig. 6.19. Relative frequency histrogram and the best fit normal curve.

This function is represented by the dotted line in Figure 6.19. The shape of the line graph conforms roughly to that of the normal curve, but it is difficult to judge the degree of adherence of the experimental results to the expected distribution.

A somewhat improved method is the comparison of the cumulative frequency of the data with the expected curve. To this end, the relative frequency of occurrence of observations not greater than x is plotted versus x. Here x is the upper limit of the pressure ranges. The expected curve is the *ogive* curve, i.e., the area under the normal curve from $-\infty$ to x:

$$F^*(x) = \frac{1}{\sigma\sqrt{2\pi}} \int_{-\infty}^{x} \exp \frac{-(x-\bar{x})^2}{2\sigma^2} \, dx$$

$F^*(x)$ is the probability of finding a value less than or equal to x in a population of measurement results. The comparison is made in Figure 6.20. It is seen that the observations, represented by the solid line, follow the cumulative probability curve (dashed line) rather closely.

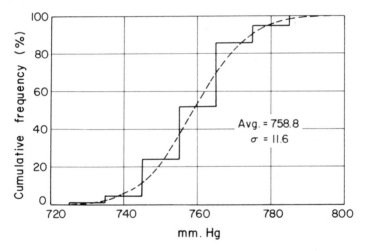

Fig. 6.20. Cumulative frequency histogram and normal distribution.

To draw the expected ogive curve for a set of data the values of the mean and the standard deviation must be calculated and from these with the aid of tables the values of $F^*(x)$. This requires much work. Therefore, special graph paper has been designed with values of $F^*(x)$ marked along one axis in percent. The distances between $F^*(x)$ and 50% are chosen to be proportional to the corresponding values of $(x - \bar{x})/\sigma$ for conveniently spaced and simple $F^*(x)$ values. This is equivalent to distorting the ordinate scale of Figure 6.20 in such a way that the cumulative probability curve (dashed line) becomes a straight line.

The type of paper referred to here is called probability graph paper or normal curve paper. Its advantage is that the data can be plotted directly without calculation of their mean and standard deviation. Since normally distributed

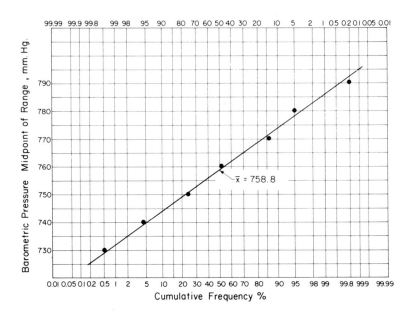

Fig. 6.21. Plot on probability paper.

variables will lie on a straight line, the plot provides a convenient test for normality of the data. The mean value can be read from the intersection of the best straight line through the data points and the 50% probability line.

In Figure 6.21 the data on barometric pressure are plotted on probability paper. From the close fit to the straight line drawn we may conclude that the data are near normally distributed. Note also that a close approximation to the mean may be read off where the line crosses a cumulative frequency of 50%. An estimate of the standard deviation may also be read off the graph. There should be a total of 34.13% of the data between the mean and +1 or –1 standard deviation from the mean. Therefore, an estimate of s^* may be read as the difference in x between 50% cumulative frequency and $50 \pm 34.13\%$ cumulative frequency. The two estimates will, of course, be the same if the data have been fitted by a straight line.

Two other types of graph paper deserve mention. On reciprocal or hyperbolic paper the abscissa is marked with distances proportional to $f(x) = 1/x$. The ordinate is divided arithmetically. Equations of the type

$$y = a + \frac{b}{x} \quad \text{or} \quad xy = ax + b$$

are represented as straight lines.

Naturally special graph paper is not available for every type of functional relationship encountered. However, by choosing the proper functions $f(x)$ and $g(y)$ of the prime variables x and y, every relation between them which contains one or two parameters can be tested by graphical means. For example, a theory of elasticity predicts the relation

$$\log y = \frac{a(p - x)}{b + x - p}$$

where p is a known constant and a and b are parameters. How can it be determined whether a series of experimantally found sets of values x, y_i are in agreement with the theory? The equation can be written as

$$\frac{1}{\log y} = \frac{b}{a(p - x)} - \frac{1}{a}$$

Now a plot is made of $g(y_i) = 1/\log y_i$ versus $f(x_i) = 1/(p - x_i)$. If the data points on this graph can be represented satisfactorily by a straight line, the data support the theory. To find the parameters a and b it is noted that the intercept of this line with the axis $f(0)$ is equal to $-(1/a)$ and that the slope of the line is b/a.

Drawings and photographs. Schematic or engineering drawings are an integral part of a technical report. They are useful for showing experimental apparatus, layout of equipment, and results. They should be used in preference to long written explanations wherever possible.

6.12 Publishing

It may some day be your pleasure or duty to have a report, technical paper, or article published in a magazine or book. The purpose of this section is to briefly outline the process from the point of view of an author.

Most journals and technical magazines have style manuals or standards of submission whose purpose is to give some uniformity to presentations. The manuals include such information as to whether engineering units of S.I. (Système Internationale, the international standard of MKS units) units are acceptable, how references are to be written, and the maximum allowable length of an article. The information for authors from the *Journal of Basic Engineering* follows.

*INFORMATION FOR AUTHORS**

Page Charges

The purpose of the Society is to disseminate technical information of permanent interest resulting in the publication of a series of Transactions quarterlies. These quarterlies are deposited, free of charge, in selected libraries throughout the world. In addition, more than 20,000 subscribers receive copies.

*Quoted with permission.

Papers of permanent interest having been selected for publication in the ASME Transactions are subject to a Page Charge of $35 per page, which will be invoiced, through the author, to the author's company, institution, or agency at the time page proofs are submitted to the author. Publication is not dependent upon payment of the page charge.

Manuscript Format

Manuscripts should be submitted in final form to the Editor. Each manuscript must be accompanied by a statement that it has not been published elsewhere nor has it been submitted for publication elsewhere. A paper which would occupy more than 8 pages of the journal may be returned to the author for abridgement.

The author should state his business connection, the title of his position, and his mailing address. A short abstract (50 to 100 words) should be included on the first page immediately preceding the introductory paragraph of the paper.

Five copies of the manuscript are required. One of these must be a carefully prepared printer's copy, typed in double spacing, on one side of the page only, with wide margins, on 8½ X 11-in. opaque white paper. Mimeographed manuscripts, if prepared with exceptional care, can be accepted provided they are completely edited.

As far as possible, all mathematical expressions should be typewritten. Greek letters and other symbols not available on the typewriter should be carefully inserted in ink. Care should be taken to distinguish between capital and lower-case letters, between zero (0) and the letter (O), between the numeral (1) and the letter (l), etc. A letter representing a vector cannot be printed with an arrow above or below it. The letter should be underscored with a single wavy line wherever it appears in the text, to designate boldface type.

A list of symbols carefully marked for the use of the editor (thus: κ, Greek l.c. kappa), if it has not been included in the body of the paper, should accompany the manuscript on a separate page.

Before preparing a manuscript the author should study printed articles in the Journal, with special attention to the form and style of mathematical expressions, tables, footnotes, references, and abstract. Numbers that identify mathematical expressions should be enclosed in parentheses. Numbers that identify references at the end of the paper should be enclosed in brackets. Care should be taken to arrange all tables and mathematical expressions in such a way that they will fit into a single column when set in type. Equations that might extend beyond the width of one column (fractions that should not be broken or long expressions enclosed in parentheses) should be rephrased to go on two or more lines within column width. Fractional powers are preferred to root signs and should always be used in more elaborate formulas. The solidus should be used instead of the horizontal lines for fractions wherever possible.

A normal paper should not exceed 8 pages in the Journal. It is about 6,000 words exclusive of drawings. A Technical Brief should not exceed 1,500 words or about two pages in the Journal inclusive of drawings.

Originals and four copies of figures must accompany the manuscript. Line drawings should not be larger than 8½ X 11 in. and should be planned for reduction to column width. Lettering should be large enough to be clearly legible when the illustration is reduced. The originals of line drawings must be in India ink on white or pale blue tracing paper or tracing cloth. Photographs of equipment or test specimens must be glossy prints and should be used sparingly. Captions for figures should be typed double-spaced and included as the last page of the manuscript. The figure number and author's name should be written in the margin or on the back of each illustration.

Titles of papers should be brief.

Authors can obtain copies of the ASME Manual MS-4, "An ASME Paper," from the Editor and are urged to do so before drafting their papers in final form.

An author is entitled to 25 preprints free of charge (in the case of two authors, 15 each; three or more authors, 10 each). Larger quantities of preprints or reprints can be ordered from Editorial Department, The American Society of Mechanical Engineers, 345 East 47th Street, New York, N.Y., 10017. Quotations will be sent on request.

Please submit manuscripts for reviews and address all communications to. . .ASME Publications Committee

The usual step-by-step procedure before an article has been submitted is given below. The paper may be invited by the editor or submitted directly.

1. Review: For articles with original scientific or technical content it is usual for the editor to have the paper reviewed for its acceptability by anonymous reviewers who render a decision to the editor and author. For journals of current interest the editor takes this responsibility himself.

2. Presentation: If the paper has been submitted to a journal of a technical society, it is possible that the report may be orally presented at a conference or symposium of the society. This may also be done independently of any publication.

3. Proofreading: When the paper has been set in type, the author will receive galley proofs. These are strips of the printed material one column wide which are to be corrected for misprints, errors, etc., by the proofreader and author.

4. Page proofs: Usually the author will receive final proofs with page numbers and page layout to check. It is very difficult for the printer to make changes at this stage and so corrections should be kept to the minimum.

5. Publication.

6. Reprints: It is common practice in the scientific and professional community for the author to obtain extra copies of his own reports and make them available, free of charge, to anyone asking for them. This practice is dying in the face of duplicating machines, but it served the purpose of keeping the author aware of other people working on the same problem.

7. Discussions: Many journals and magazines print discussions on reports and articles previously published. Much of this discussion is critical of the original work—often highly so. Science and technology proceed upon the basis of accepted theories and experimental results. It is very important that all contributions be open to question at all times. Here are some examples:

> *Science Letters* column—Nov. 29, 1968.

> 1. Though Bardech and this letter point out some of the many technical problems which remain, I feel there is a bright future for aquaculture as a source of food and other products needed by man.*

> *Science Letters* column—Nov. 5, 1968.

> 2. . . .Fish may have many local names but to transplant Boston mackeral from New England to the Pacific Ocean and name them Pacific salmon is jolting. . .*

> *Applied Mechanics Reviews,* Sept. 1968, p. 953.

> 3. I am surprised at the small-scale apparatus used for these experiments, as it could induce severe secondary effect. . . .*

> Letter in *Mechanical Engineering,* Feb. 1971.

> 4. The difficulties in planning a consistent anti-pollution program are nowhere more evident than in. . . .On page 16 we read that "materials that are incompatibly reclaimable should not be used together," and a suggestion for "the use of invisible ink to facilitate the recovery of a newsprint." Then on p. 19 we have another idea: the stapling of newspapers to avoid the littered condition which results when a newspaper comes apart of a wind-blown street." Washable staples perhaps.*

6.13 Exercises

6.1. Write an abstract of any chapter of any textbook which you are using. Remember to be specific in naming the contents of the chapter.

6.2. Obtain the population statistics for the previous 10 years for a city or town with which you are familiar. Using the techniques of this chapter, attempt to illustrate whether or not this population follows a functional form.

6.3. In your library you will find government-prepared statistical summaries of population, economic and financial variables, and census material for your

*Quoted with permission.

country, state, province, county, etc. Choose some table of data of interest to you and present it in at least three different forms. Attempt to illustrate that different methods of presentation illustrate different aspects of the data used.

6.4. Use the technique of probability paper to solve Exercises 4.2(d) and (e) and 4.6.

6.5. Results from a tensile test on 36 steel cables chosen at random from a certain mill were grouped as follows:

Tensile strength (kip/sq in.)	155–165	165–175	175–185	185–195	195–205	205–215	215–225
Frequency	1	4	7	11	9	3	1

(a) Find the mean and standard deviation of the tensile strength of the cable.

(b) Plot on normal probability paper the tensile strength versus fractional cumulative frequency. Draw "by eye" a straight line through the points and obtain the mean and standard deviation from the graph, and observe whether the results appear to follow the normal distribution.

(c) Estimate the probability of obtaining a random measurement which has a deviation from the mean of between 10 and 20 kip/sq in.

(d) Calculate the range in which we would expect the mean of the population to fall with a probability of 95%.

6.6. Refer to Figure 6.16. What is the composition of the mixture represented by point Q. Read from the diagram whether the following mixtures are homogeneous at 30°C and 1 atm pressure:

(a) Phenol : acetone : water = 35:40:25 wt %.

(b) Phenol : acetone : water = 40:40:20 wt %.

(c) Equal weights of the three components.

6.14 Suggestions for Further Reading

Blicq, R. S., *Technically Write!*, Prentice-Hall, Inc., Englewood Cliffs, N.J., 1972.

Style Manual, 3rd ed., American Institute of Physics, New York, 1961.

Wirkus, T. E., and Erickson, H. P., *Communication and the Technical Man,* Prentice-Hall, Inc., Englewood Cliffs, N. J., 1972.

7

Example Reports

This chapter contains four technical reports, two from published literature and two produced by students working on undergraduate projects. These reports are not necessarily intended as models but are, however, intended to illustrate some of the variety of methods available for reporting technical work. They are as follows:

Some brief comments on these reports are given on page 162.

Variation in Quality Between Brands of Table Tennis Balls

Table of Contents

Abstract

A project has been performed to measure the weight, the diameter, the roundness (in terms of the diameter), and the coefficient of restitution of four brands of ping-pong balls. In this project, five balls of each brand were tested.

From the test results it is shown that the bouncing qualities of two brands are high. The weight and the diameter of almost all the balls are within the standards set by the International Table Tennis Federation. Brand D is the worst of the four brands being tested. This is shown by its weight, which is far from the standard, and its comparatively poor bounceability. In between these two extremes is Brand C, whose qualities are neither as good nor as well controlled as Brands A and B. Nevertheless, test results show that a slight majority of its balls are within international standards and that its use for general recreational purposes can be justified.

1

Variation in Quality Between Brands of Table Tennis Balls

<u>Introduction</u>

Table tennis has always been the most popular sport in China. It is a good game for recreational purposes, and many families have set up their own facilities for play. With the increasing popularity of the game, different brands of balls which vary in quality and price have become available on the market. It is important, therefore, for a player to know which brand is best suited to his own use. A knowledge of the weight, the roundness, and the bounceability of each brand of ball is therefore useful in making the judgment.

The weight and the diameter of the ball can be measured with the use of an electric balance and a vernier caliper, respectively. The coefficient of restitution is found by relating it, mathematically, to the height of release of a ball and the time the ball takes to come to a full rest.

The objective is to establish the weight, diameter, and coefficient of restitution of four brands of table tennis balls. The results obtained through experiment will be compared with the established standards of the International Table Tennis Federation. (The standards can be found in Appendix I.) Through this comparison, the choice of the different brands for any purpose can be determined.

2

Procedure

Four common brands of table tennis balls, which we shall title Brands A, B, C and D, are being tested in this project. The weight of each ball is found by weighing it on an electric balance (± 0.0001 g). The diameter of each brand is measured in the following way: Imagine that there are three mutually perpendicular axes with their origin at the center of the ball. Measurement of the diameter is taken along these three axes. With five balls of each kind, the arithmetic mean of the 15 measurements gives the mean diameter of that kind. Also, since the ball is not perfectly spherical, the standard deviation of each kind is a reasonable estimate of the degree of roundness. Small standard deviation implies that the ball is more "round" than a ball with a larger standard deviation.

The coefficient of restitution is defined as the ratio between the velocity of the object immediately after the impact to the velocity of approach, that is,

$$e = \frac{\text{Instantaneous velocity after impact}}{\text{Instantaneous velocity before impact}}$$

where e = coefficient of restitution. The method used involved the measurement of the height at which the ball is let off and the time the ball takes to come to a full rest. The time, initial height, and the coefficient of restitution are related in the following equation:

$$e = \frac{T - \sqrt{2h_0/g}}{T + \sqrt{2h_0/g}}$$

where   e = coefficient of restitution

T = total time

h_0 = initial height

g = gravitational acceleration

A complete development of the equation is attached in Appendix II.

3

The initial height was picked as 80 cm above the surface of a table tennis table. This was chosen as a compromise after a series of experiments using different initial heights was done. The lower the initial height, the shorter the time and the larger the chance of experimental error. The higher the height, the larger is the effect of air resistance to the ball. It will tend to slow down the ball.

Observations

With the results at hand, the mean and the standard deviation are calculated and presented in Table I. Table II shows the percentage of the balls that are within standards with respect to its different qualities.

The mean diameter of these four brands are fairly close to each other. Brand A has the smallest standard deviation and Brand D the largest. This means that Brand A is closest to spherical among the four brands tested. The standard has upper and lower tolerance limits, and all these balls complied to the standard.

The weight of Brands A and B are quite acceptable. Statistics show that 86.9 and 90.8% of the balls are likely to be within the standard. Brand C is slightly heavier, and only 54.5% are within the standard. The weight of Brand D is nowhere near the standard. It is far too light for a ping-pong ball.

Surprisingly, Brand D is the only one that gives a coefficient of restitution that is within the standard. The other three brands seem to have too high a coefficient. If one were to draw the conclusion that Brand D has the truest bounce, one would be wrong. Other factors that have to be considered before any conclusions can be drawn will be discussed in the following section.

The results of the tests are summarized in Figures 1 and 2.

4

Comparison of Theory and Results

Considering the variables that might offset the results, it is clear that measurement of the weight of the ball presents the least variation. The only likely source of error is in the calibration and functioning of the electric balance itself.

All four brands of balls are spherical to the eye. The reading of the diameter can be influenced by thermal expansion due to the hand or by deformation by the calipers; however, these effects were assumed to be small and were neglected.

Experimental results show that Brands A, B, and C give better bounce, but they seemed to give too high a coefficient of restitution. The value of e depends not only on the nature of the ball, but also on the nature of the other body with which the ball collided (in our case it is the table tennis table). A ball will not have a high value of e if it is dropped onto a pool of sand. In our case energy lost during impact was not as much as it should be. It is impor- tant to remember that there are standards governing the table top as well. A test of the nature of the table top is beyond the scope of this project.

Conclusions

From the results it can be concluded that Brands A and B are basically better balls. Brand C is suitable for recreational use, and Brand D is not suitable for any purpose.

5

Acknowledgments

I would like to take the opportunity here to thank Mr. John Doe for his help and guidance in this project.

Bibliography

1. Fender, D. H., General Physics and Sound, The English Universities Press Ltd., London, 1970.
2. White, Jess R., Sports Rules Encyclopedia, The National Press, Palo Alto, Calif., 1972.

6

Table I

Summary of Data in Terms of Mean and Standard Deviations

Quality-Tested Brands		Weight (g)	Diameter (in.)	Coefficient of Restitution
A	Mean	1.491	2.4406	0.901
	Standard deviation	0.003	0.0363	0.0031
B	Mean	1.477	2.4688	0.895
	Standard deviation	0.005	0.0505	0.0017
C	Mean	1.483	2.5410	0.892
	Standard deviation	0.003	0.1270	0.0045
D	Mean	1.488	1.6784	0.836
	Standard deviation	0.007	0.0997	0.0025
	Standard	1.43-1.51	2.40-2.66	0.816-0.866

Table II

Percentage of Balls of Each Brand that Are Within Set Standards

Quality-Tested Brands	Weight (%)	Diameter (%)	Coefficient of Restitution
A	86.90	100	—*
B	90.84	100	—
C	54.37	100	—
D	0.00	100	—

*The percentage for coefficient of restitution is not presented. An explanation is given under the heading "Comparison of Theory and Results."

7

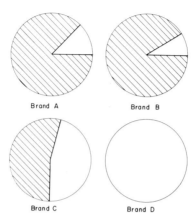

Fig. 1 Comparison of brands by weight; percentage of ball (represented by shaded area) with respect to weight that are within the standard.

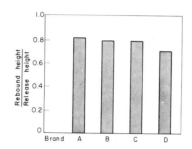

Fig. 2 Comparison of brands by rebound.

8

Appendix I

The standard of the table tennis balls set by the International Table Tennis Federation is given in a book entitled Sports Rules Encyclopedia. It states that

1. The ball shall be optically spherical. It shall be between 4 1/2 and 4 3/4 in. inclusive in circumference. It shall weigh not less than 37 grains nor more than 41 grains (2.40 to 2.66 g).

2. The table shall yield a uniform bounce of 8–9 in. when a ball is dropped from a height of 12 in. above the surface of the table. (That is, the coefficient of restitution can vary from 0.816 to 0.866.)

Appendix II

This appendix contains a derivation of the formula used to calculate e and is not presented here.

9

Abstract

Deformability of Elastic Bands

A project has been performed to measure the load-elongation characteristics of two brands of rubber bands (Brands A and B). In this project, the elasticity, maximum elongation, and load of six bands from each brand were tested.

From the test results, it can be concluded that Brand B rubber bands stretch more easily and can be stretched further before failure occurs. This is shown by the modulus of elasticity and maximum elongation of the rubber bands. Further, the variation of these values is smaller, as shown by the standard deviation of the test results. Finally, both brands broke under a similar maximum load of 15 lb.

1

Deformability of Elastic Bands

Introduction

Elastic bands find many uses in industry, commerce, and households. It is desired to measure the elasticity, maximum elongation, and maximum strength of the elastic bands.

The measurement will be performed by stretching the elastic bands by hanging weights from the bands and measuring the elongation and rupture strength.

The objective is to establish the elasticity, maximum elongation, and maximum strength of two brands of elastic bands. The results will be helpful in the choosing of the best brand for any particular purpose.

2

Two common brands of elastic bands, called here Brands A and B, have been chosen for this measurement project. These bands have a 6-in. circumference in the unstretched condition. More detail is given in Table 1.

Table 1

	Brand	
	A	B
Folded length, in.	3.0	3.0
Width, in.	0.113	0.113
Thickness, in.	0.041	0.041
Color	Multiple	Brown
Number tested	6	6

The equipment used in this measuring project was composed of a retort stand, hanging arrangement, calibrated weights, and a 6-ft steel tape marked off in inches and tenths of inches. A sketch of the equipment is shown in Figure 1.

The elastic bands were stretched by hanging measured weights on the hanger. At the start, the weights were added in 0.4-lb increments to establish the initial linear load-deformation characteristic. Then the load was added in 2.0-lb increments, and, finally, close to the failure strength, the loads were added again in 0.4-lb increments to establish the failure load. These increments were considered reasonable in view of the large variation in the failure load of the bands.

Six bands from each of these two brands of elastic bands were tested. These were intended as a random sample from the larger population.

3

Observations

The results of the tests are summarized in Table 2. Further typical load-elongation curves are shown in Figure 2. The results will be discussed under two headings: elasticity and failure.

Elasticity:

To express the ease of stretch of the bands, consider the curves in Figure 2. The load-elongation curves are initially linear to about a 3-lb load and then they curve. Further, the Brand B line stretches more than the Brand A under the same load. The elasticity of a material is defined by the linear portion of the curves, and it is defined by the elastic modulus. The elastic modulus is calculated from the following relationship:

$$E = \frac{\sigma}{\epsilon} = \frac{P/A}{\Delta 1/\ell} \tag{1}$$

where E = modulus of elasticity, psi

σ = stress, psi

ϵ = strain, in./in.

P = load, lb

A = area, in.^{2}

$\Delta 1$ = elongation, in.

1 = initial length, in.

Using equation (1), the elastic modulus at the 3-lb load state was calculated for each test. The results are summarized in Table 2. Further, the mean and standard deviation of the results of the two brands are shown in the table.

4

Considering the results in Table 2, it can be observed that Brand A rubber bands have a higher mean elastic modulus and a larger standard deviation than Brand B. This means that Brand A rubber bands are stiffer and stretch less under a given load and that there is a larger variation in the elasticity of these brands, whereas Brand B bands stretch more easily and have a more uniform elastic property.

Failure:

Failure of the elastic bands can be defined by either the maximum elongation they will tolerate or the maximum load they can withstand.

The maximum elongations and loads for the tests are given in summary form in Table 2.

Considering the maximum elongation of the two brands, it can be concluded that Brand B rubber bands are the better product. From Table 2 it can be seen that the mean maximum elongation is 23.1 in. (versus 22.2 in. for Brand A) and further that the variation of the maximum elongation is smaller as shown by the standard deviations (0.9 in. for Brand B versus 1.6 in. for Brand A).

Finally, the elastic bands can be compared by the maximum load they can withstand. From the results given in Table 2 it can be observed that there is not a great difference in the means of the maximum load or the standard deviations. The mean maximum load for Brand A was found to be 15.4 lb, and for Brand B, 15.3 lb. Therefore it can be concluded that they fail under a similar weight of approximately 15.3 lb.

5

Table 2

	Brand			
	A		B	
	Mean	Standard Deviation	Mean	Standard Deviation
Elastic modulus, psi	66.3	7.1	57.3	2.6
Maximum elongation, in.	22.2	1.6	23.1	0.9
Maximum load, lb	15.4	1.7	15.3	2.2

Conclusions

From the measuring project performed on six bands of each brand of rubber bands, it can be concluded that the Brand B rubber bands are of higher quality. This conclusion is based on the following observations:

1. Brand B bands stretch easier as is shown by the lower modulus of elasticity.
2. Brand B bands can withstand a larger elongation.
3. A smaller variation of values 1 and 2 was found with the Brand B bands. Numerical values of the standard deviations are shown in Table 2.

Finally, it can be concluded that they have a similar maximum strength, as both failed under approximately 15-lb loads.

6

152

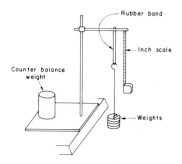

Fig. 1 Testing arrangement

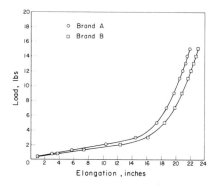

Fig. 2 Typical load-elongation curves for rubber bands.

7

Procedure for Simple, High Precision Density
Determinations of Air for Buoyancy Corrections

J. Brulmans and H. L. Eschbach,

Bureau Central de Mesures Nucleaires, Geel, Belgium.
(Received June 8, 1970 and in final form July 17, 1970.)

The most important correction that enters very accurate mass determinations is that for buoyancy forces exerted on the object to be weighed. These corrections depend on the volume V of the object and the density of air ρ_A. As is well known, major changes in the density of air occur due to variation in the temperature, pressure, and relative humidity, and it is of high importance to measure these ambient parameters with sufficient accuracy. [Changes may also occur due to variation in the composition of the air (Kohlrausch[1]), but for the experiments to be described it may be assumed that no significant change in the composition of the air took place.] Bowman and Schoonover[2] (1967) have derived a formula which allows us to calculate with adequately small error the density ρ_A of air (mg/cm^{-3}) when the barometric pressure B (Torr), the temperature t_A ($^{\circ}$C), and the relative humidity H (%) are known. For the temperature range between 20 and 30°C they derived for the density of air

$$\rho_A = \frac{0.464554B - H(0.00252t_A - 0.020582)}{t_A + 273.16} \tag{1}$$

[1] F. Kohlrausch, Praktische Physik (B. G. Teubner, Stuttgart, 1955).

[2] H. A. Bowman and R. M. Schoonover, J. Res. Nat. Bur. Stand. 71C, 179 (1967).

1

Table 1

Dimension and Properties of the Aluminum Capsule of Figure 1

Volume at 20°C	$V_{A\ell}(cm^3)$	$69.448_3 \pm 0.0005$
Mass	$M_{A\ell}$ (g)	$23.147720 \pm 5 \times 10^{-6}$
Coefficient of thermal expansion	$(°C^{-1})$	77.9×10^{-6}
Compressibility	$Torr^{-1}$	9.7×10^{-7}

In many applications it is difficult and time consuming to measure the three parameters with sufficient precision and it was, therefore, tried to extend a method indicated already by Bowman and Schoonover[2] in order to measure the density of air directly. To this end a tight capsule made of aluminum with a volume $\underline{V}_{A\ell}$ and a true mass $\underline{M}_{A\ell}$ was compared with a weight of stainless steel of nearly equal mass $\underline{M}_s$ and having a volume $\underline{V}_s$. If $\underline{G}_{A\ell}$ and $\underline{G}_s$ are the apparent weights of the aluminum capsule and the stainless steel weight, respectively, the following expression can be derived:

$$\frac{M_{A\ell} - G_{A\ell}}{\rho A} - \frac{M_s - G_s}{\rho A} + V_A - V_s \qquad (2)$$

or, with obvious abbreviations,

$$\Delta G = -\Delta V_{\rho_{A\ell}} + \Delta M \qquad (3)$$

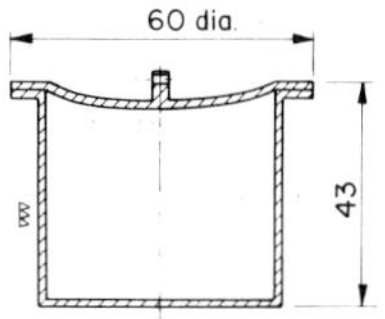

Fig. 1 Aluminum capsule filled with helium and tightened by cold welding; dimensions are given in millimeters.

2

Thus, the difference of apparent weight ΔG is a linear function of the density of air ρ_A. The capsule used (Figure 1) was similar to those developed for the encapsulation of Ge(Li)-detector as described by Meyer et al.[3] After cleaning, degassing, and evacuating the capsule it was filled with helium to about atmospheric pressure and sealed by cold pressure welding. Thus organic sealants which might give difficulties due to sorption or diffusion processes are avoided. The volume of the capsule, its coefficient of thermal expansion, and its compressibility were determined by hydrostatic weighing. For the evaluation of the thermal expansion the volume was measured at 20 and 30°C. The compressibility was determined by placing the capsule at a depth of 13 and 33 cm with respect to the surface of the water bath. The measured values are incorporated in Table 1.

From the figures in the table it can be seen that for measurements at temperatures other than 20°C correction of the volume due to thermal expansion has to be applied whereas the compression of the capsule due to various barometric pressures can be neglected.

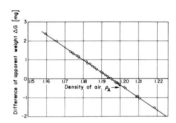

Fig. 2   Testing of air ρ_4 versus the difference of apparent weight of the aluminum capsule.

The differences in apparent weight were measured during a period of more than 1 yr at very differing atmospheric pressures. The corresponding air densities were calculated according to Eq. (1) and plotted against ΔG

[3]H. Meyer, H. L. Eschbach, W. Nagel, and E. W. Kruidhof, EURATOM, 4063.e (1968).

3

(Figure 2). The straight line was calculated from a least squares fit through the measured points. This gives a very good agreement (within the errors quoted) with the line which can be calculated according to Eq. (3). From these experiments it can be concluded that direct measurements of air density can be carried out by determining the changes of the apparent weight of a hollow aluminum cylinder with a volume of about 70 cm^3. The mass of displaced air corresponding to this volume is about 80 mg. Since the experiments were carried out on a microbalance with a precision of $\pm 5 \times 10^{-6}$ g, the density of air can be determined with a precision $\pm 7.5 \times 10^{-8}$ g-cm^{-3}.

The method described is not limited to weighings with microbalances only. As the buoyancy force on the capsule with its average density of about 0.33 g-cm^{-3} is much higher than on most other objects weighed the method can be applied on all types of analytical balances with the same capsule. The only limitation of the accuracy is imposed by the weighing accuracy of the balance used. It should be noted that this method can easily be adapted to cases where densities of gases other than air have to be measured, for instance, when weighing under inert gas in a glovebox.

We wish to thank Mr. H. Moret for many helpful discussions. The assistance of F. Hendrickx, who carried out the weighings is gratefully acknowledged.

4

From Review of Scientific Instruments, 41, No. 11, 1970, p. 1680. Quoted with permission.

7.4 Old Faithful: A Physical Model

Abstract

Old Faithful: A Physical Model

The recent confirmation of a prediction that relates the duration of eruption to time between eruptions suggests a physical model of the internal cavity of Old Faithful.

1

Old Faithful: A Physical Model*

Rinehart (1) has proposed that Old Faithful exhibits two modes of
eruption. One mode (a long interval between eruptions) consists of a 20-
to 30-min quiet period and is concluded by a brief series of long-period
ground movements that are followed by general seismic activity (weak
tremors and strong pulses) that continue up to the time of eruption. The
second mode (a short interval between eruptions) is characterized by
immediate seismic activity, consisting of both weak tremors and strong
pulses and occasionally including the introductory long-duration movements.
This activity continues up to eruption. There is no observable relationship
between long and short intervals between eruptions. Old Faithful has also
been observed to exhibit a range of from 1.5 to 4.5 min in the duration of
water play.

On the basis of these data (1), I proposed (2) that Old Faithful consists
of a single cavity that occasionally was incompletely emptied in an eruption.
If we make the assumption that seismic activity is related to the boiling
activity in the geyser cavity, we can explain (1) the immediate onset of
seismic activity (in cases of a short interval between eruptions) and (2) the
existence of short-interval eruptions, by hypothesizing that this mode of
eruption results when the prior eruption incompletely evacuates the cavity,
and hence the cavity remains partially full of hot liquid. Therefore, there
would be immediate boiling activity, and less time would be necessary to
reach the critical point for a second eruption.

In cases of complete eruption, the cavity is essentially empty, and
there would follow a longer interval preceding the next eruption. The first
part of this period would be quiet because there would not already be a
mass of hot boiling liquid in the cavity. On the basis of the above rationale,
I proposed that the relation between duration of water play (this should provide
an estimate of how completely the cavity is emptied) and the length of the

2

Science, 160, No. 3831, May 31, 1968, p. 989-990. Quoted with permission.

interval to the subsequent eruption should be examined, and I hypothesized that a brief duration of water play (incomplete emptying) would be associated with a brief interval to the following eruption.

 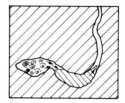

Fig. 1   Cavity Shape (Left) Cavity filled to point at which the surging motion of the water in the U portion of the cavity (which results from the venting of steam from the back portion) would produce noticeable ground tremors and pulses. (Right) Water splashing over into the hot, dry, back portion of the cavity flashes into steam and blows the liquid up and out of the U portion of the cavity.

This hypothesis was tested by Rinehart and confirmed (3); its confirmation suggests that there is a physical model for the geyser cavity (Figure 1). During the quiet portion of a long interval between eruptions, the lower U part of the cavity would slowly fill and come to a boil. Seismic activity would result when the mass of boiling water was great enough to seal off the U portion of the cavity. This would mean that venting steam from the back half of the cavity would have to blow out through the water mass, producing detectable vibrations. An eruption would take place when the U portion of the cavity was sufficiently full to splash a quantity of water over into the hot, dry, back half of the cavity. The water would immediately flash boil to steam, forcing the water out of the U section of the cavity. A large splash would generate a large volume of steam and blow the cavity clean. A smaller splash would result is a partial evacuation of the cavity. If, in the latter event, sufficient water remained in the cavity to seal off the U portion of the cavity, new seismic activity would start immediately.

3

Data on the volume of water erupted would provide an additional test of the model. Because the geyser opening is a constant, a measure of the velocity of the escaping liquid in addition to the duration of water play would provide a good estimate of the volume ejected. On the basis of the above model, one would predict that eruptions which precede a long interval (complete emptying) would involve longer water play, a greater total volume ejected, and also a greater water eruption velocity. (It has been hypothesized above that a greater volume of steam is generated in these cases.) Therefore, the internal pressure should be greater, and the eruption velocity greater. Although an additional test should be made, the present data seem to lend adequate support to the proposed model to warrant its temporary adoption.

Fred Geis, Jr.

Longfellow Hall, Harvard University,
Cambridge, Massachusetts 02138.

References

1. J. S. Rinehart, Science 150, 494 (1965).
2. F. Geis, Ibid. 151, 223 (1966).
3. F. S. Rinehart, personal communication, April 14, 1968.

4

Comments on the Report on Table Tennis Ball Quality. The abstract of this report is rather long compared to the extent of the report itself, but notice that a quick review of the abstract tells you quite clearly that there are two recommended brands and one nonrecommended brand and gives the basis for the statement.

The introduction makes a very clear statement of objectives and gives the source of comparison of the brands (International Table Tennis Federation Rules). The procedure is clear and simple, all terms are defined, and possible errors are outlined. The rest of the report is devoted to a discussion of these results. Notice also how, to a certain extent, diagrams may be read independently of the text.

Comments on Deformability of Elastic Bands. Notice that the abstract is completely self-contained. The statement of objectives is in the second sentence of the introduction and the method in the next sentence. Table 1 and Figure 1 summarize a considerable quantity of detail which would have been much too long if written in straight prose. Notice that only six bands were tested. This is sufficient in this case since large samples are a poor way to achieve precision. Formula (1) should have a reference. In the discussion of failure, the author notes that failure can be defined in more than one way and tabulates both. Note that the conclusions can be read more or less on their own. This is true also of the figures.

Comments on the Report on Density Determinations of Air for Buoyancy Corrections. The title is long but necessary in this report because of the highly specialized application of the work done. The date under the authors' name establishes precedence in the event of two similar discoveries at a similar time. There is no abstract as such, since this article is so short as to be effectively an abstract in itself. The statement of the problem is given in the first sentence below Table 1. Note that the mathematical steps between Equations (2) and (3) are omitted in the interest of brevity.

Comments on Old Faithful: A Physical Model. Technical reports and scientific papers are by no means always mathematical or numerical in nature. This report is entirely descriptive. Note that despite the structure of the report, it is still similar to the others, with an abstract, an introduction, a basic proposition, a discussion, and conclusions.

Appendix I

Tables

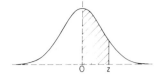

Table 1.1.   Areas Under the Standard Normal Curve from 0 to z

z	0	1	2	3	4	5	6	7	8	9
0.0	.0000	.0040	.0080	.0120	.0160	.0199	.0239	.0279	.0319	.0359
0.1	.0398	.0438	.0478	.0517	.0557	.0596	.0636	.0675	.0714	.0754
0.2	.0793	.0832	.0871	.0910	.0948	.0987	.1026	.1064	.1103	.1141
0.3	.1179	.1217	.1255	.1293	.1331	.1368	.1406	.1443	.1480	.1517
0.4	.1554	.1591	.1628	.1664	.1700	.1736	.1772	.1808	.1844	.1879
0.5	.1915	.1950	.1985	.2019	.2054	.2088	.2123	.2157	.2190	.2224
0.6	.2258	.2291	.2324	.2357	.2389	.2422	.2454	.2486	.2518	.2549
0.7	.2580	.2612	.2642	.2673	.2704	.2734	.2764	.2794	.2823	.2852
0.8	.2881	.2910	.2939	.2967	.2996	.3023	.3051	.3078	.3106	.3133
0.9	.3159	.3186	.3212	.3238	.3264	.3289	.3315	.3340	.3365	.3389
1.0	.3413	.3438	.3461	.3485	.3508	.3531	.3554	.3577	.3599	.3621
1.1	.3643	.3665	.3686	.3708	.3729	.3749	.3770	.3790	.3810	.3830
1.2	.3849	.3869	.3888	.3907	.3925	.3944	.3962	.3980	.3997	.4015
1.3	.4032	.4049	.4066	.4082	.4099	.4115	.4131	.4147	.4162	.4177
1.4	.4192	.4207	.4222	.4236	.4251	.4265	.4279	.4292	.4306	.4319
1.5	.4332	.4345	.4357	.4370	.4382	.4394	.4406	.4418	.4429	.4441
1.6	.4452	.4463	.4474	.4484	.4495	.4505	.4515	.4525	.4535	.4545
1.7	.4554	.4564	.4573	.4582	.4591	.4599	.4608	.4616	.4625	.4633
1.8	.4641	.4649	.4656	.4664	.4671	.4678	.4686	.4693	.4699	.4706
1.9	.4713	.4719	.4726	.4732	.4738	.4744	.4750	.4756	.4761	.4767
2.0	.4772	.4778	.4783	.4788	.4793	.4798	.4803	.4808	.4812	.4817
2.1	.4821	.4826	.4830	.4834	.4838	.4842	.4846	.4850	.4854	.4857
2.2	.4861	.4864	.4868	.4871	.4875	.4878	.4881	.4884	.4887	.4890
2.3	.4893	.4896	.4898	.4901	.4904	.4906	.4909	.4911	.4913	.4916
2.4	.4918	.4920	.4922	.4925	.4927	.4929	.4931	.4932	.4934	.4936
2.5	.4938	.4940	.4941	.4943	.4945	.4946	.4948	.4949	.4951	.4952
2.6	.4953	.4955	.4956	.4957	.4959	.4960	.4961	.4962	.4963	.4964
2.7	.4965	.4966	.4967	.4968	.4969	.4970	.4971	.4972	.4973	.4974
2.8	.4974	.4975	.4976	.4977	.4977	.4978	.4979	.4979	.4980	.4981
2.9	.4981	.4982	.4982	.4983	.4984	.4984	.4985	.4985	.4986	.4986
3.0	.4987	.4987	.4987	.4988	.4988	.4989	.4989	.4989	.4990	.4990
3.1	.4990	.4991	.4991	.4991	.4992	.4992	.4992	.4992	.4993	.4993
3.2	.4993	.4993	.4994	.4994	.4994	.4994	.4994	.4995	.4995	.4995
3.3	.4995	.4995	.4995	.4996	.4996	.4996	.4996	.4996	.4996	.4997
3.4	.4997	.4997	.4997	.4997	.4997	.4997	.4997	.4997	.4997	.4998
3.5	.4998	.4998	.4998	.4998	.4998	.4998	.4998	.4998	.4998	.4998
3.6	.4998	.4998	.4999	.4999	.4999	.4999	.4999	.4999	.4999	.4999
3.7	.4999	.4999	.4999	.4999	.4999	.4999	.4999	.4999	.4999	.4999
3.8	.4999	.4999	.4999	.4999	.4999	.4999	.4999	.4999	.4999	.4999
3.9	.5000	.5000	.5000	.5000	.5000	.5000	.5000	.5000	.5000	.5000

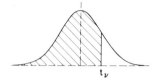

Table I.2.   Student's t Distribution for ν Degrees of Freedom (One Side)

ν	$t_{.995}$	$t_{.99}$	$t_{.975}$	$t_{.95}$	$t_{.90}$	$t_{.80}$	$t_{.75}$	$t_{.70}$	$t_{.60}$	$t_{.55}$
1	63.66	31.82	12.71	6.31	3.08	1.376	1.000	.727	.325	.158
2	9.92	6.96	4.30	2.92	1.89	1.061	.816	.617	.289	.142
3	5.84	4.54	3.18	2.35	1.64	.978	.765	.584	.277	.137
4	4.60	3.75	2.78	2.13	1.53	.941	.741	.569	.271	.134
5	4.03	3.36	2.57	2.02	1.48	.920	.727	.559	.267	.132
6	3.71	3.14	2.45	1.94	1.44	.906	.718	.553	.265	.131
7	3.50	3.00	2.36	1.90	1.42	.896	.711	.549	.263	.130
8	3.36	2.90	2.31	1.86	1.40	.889	.706	.546	.262	.130
9	3.25	2.82	2.26	1.83	1.38	.883	.703	.543	.261	.129
10	3.17	2.76	2.23	1.81	1.37	.879	.700	.542	.260	.129
11	3.11	2.72	2.20	1.80	1.36	.876	.697	.540	.260	.129
12	3.06	2.68	2.18	1.78	1.36	.873	.695	.539	.259	.128
13	3.01	2.65	2.16	1.77	1.35	.870	.694	.538	.259	.128
14	2.98	2.62	2.14	1.76	1.34	.868	.692	.537	.258	.128
15	2.95	2.60	2.13	1.75	1.34	.866	.691	.536	.258	.128
16	2.92	2.58	2.12	1.75	1.34	.865	.690	.535	.258	.128
17	2.90	2.57	2.11	1.74	1.33	.863	.689	.534	.257	.128
18	2.88	2.55	2.10	1.73	1.33	.862	.688	.534	.257	.127
19	2.86	2.54	2.09	1.73	1.33	.861	.688	.533	.257	.127
20	2.84	2.53	2.09	1.72	1.32	.860	.687	.533	.257	.127
21	2.83	2.52	2.08	1.72	1.32	.859	.686	.532	.257	.127
22	2.82	2.51	2.07	1.72	1.32	.858	.686	.532	.256	.127
23	2.81	2.50	2.07	1.71	1.32	.858	.685	.532	.256	.127
24	2.80	2.49	2.06	1.71	1.32	.857	.685	.531	.256	.127
25	2.79	2.48	2.06	1.71	1.32	.856	.684	.531	.256	.127
26	2.78	2.48	2.06	1.71	1.32	.856	.684	.531	.256	.127
27	2.77	2.47	2.05	1.70	1.31	.855	.684	.531	.256	.127
28	2.76	2.47	2.05	1.70	1.31	.855	.683	.530	.256	.127
29	2.76	2.46	2.04	1.70	1.31	.854	.683	.530	.256	.127
30	2.75	2.46	2.04	1.70	1.31	.854	.683	.530	.256	.127
40	2.70	2.42	2.02	1.68	1.30	.851	.681	.529	.255	.126
60	2.66	2.39	2.00	1.67	1.30	.848	.679	.527	.254	.126
120	2.62	2.36	1.98	1.66	1.29	.845	.677	.526	.254	.126
∞	2.58	2.33	1.96	1.65	1.28	.842	.674	.524	.253	.126

Appendix II

Proof of Bessel's Correction

A sample of size n drawn from a population with a mean μ and standard deviation σ will have the following properties:

Let x_i be an observation in the sample. Then

$$x_i - \mu = (x_i - \bar{x}) + (\bar{x} - \mu)$$
$$= (x_i - \bar{x}) - \epsilon \tag{1}$$

$\epsilon = \mu - \bar{x}$ is the deviation of the sample mean, $\bar{x}$ from the "true" value of the "error" in $\bar{x}$. Squaring equation (1), we obtain

$$(x_i - \mu)^2 = (x_i - \bar{x})^2 + \epsilon^2 - 2\epsilon(x_i - \bar{x}) \tag{2}$$

Sum for i from 1 to n, and obtain

$$\Sigma (x_i - \mu)^2 = \Sigma (x_i - \bar{x})^2 + n\epsilon^2 - 2\epsilon \Sigma (x_i - \bar{x}) \tag{3}$$

This will apply to all the observations in the sample:

$$\Sigma (x_i - \bar{x}) = 0 \quad \text{by definition of } \bar{x}$$

Therefore,

$$\Sigma (x_i - \mu)^2 = \Sigma (x_i - \bar{x})^2 + n\epsilon^2 \tag{4}$$

Repeat this calculation for a large number of samples. Then, the mean value of the left-hand side of equation (4) will (by the definition of σ^2) tend toward $n\sigma^2$. Similarly, the mean value of $n\epsilon^2 = n(\mu - \bar{x})^2$ will tend toward n times the

variance of $\bar{x}$, since ϵ represents the deviation of the sample mean from the population mean. Thus,

$$n\epsilon^2 \to n\left(\frac{\sigma^2}{n}\right)$$

Therefore, putting these two limits into equation (4),

$$n\sigma^2 \to \Sigma\,(x_i - \bar{x})^2 + \sigma^2$$

or

$$\Sigma\,(x_i - \bar{x})^2 \to (n - 1)\sigma^2$$

Thus,

$$\frac{\Sigma\,(x_i - \bar{x})^2}{n - 1} \to \sigma^2 \quad \text{Q.E.D.} \tag{5}$$

In other words, for a *large number* of random samples, the mean value of $[\Sigma\,(x_i - \bar{x})^2]/(n - 1)$ tends toward σ^2; that is, it is an unbiased estimate of the variance of the population. The estimate is denoted by s^2. Thus,

$$s^2 = \frac{\Sigma\,(x_i - \bar{x})^2}{n - 1}$$

Since the variance of the sample (taken as a finite population with a mean $\bar{x}$) σ^2 is given by

$$\sigma^2 = \frac{\Sigma\,(x_i - \bar{x})^2}{n}$$

Bessel's correction is

$$\frac{s^2}{\sigma^2} = \frac{n}{n - 1}$$

or the square root can also be called Bessel's correction:

$$\frac{s}{\sigma} = \sqrt{\frac{n}{n - 1}}$$

Appendix III

Suggestions for Projects

and Discussion Topics

The projects suggested by the titles in this list may be used to obtain practice in the techniques presented in this book. It is recommended that a project be done in five distinct stages:

1. Choose a topic. If possible choose a topic of personal interest.
2. Prepare a detailed experimental plan. Chapter 2 should provide guidance.
3. Perform the experiment or measurement.
4. Prepare a draft report for criticism by a superior.
5. Submit a final report on the results of the work.

List of Typical Projects Performed by Undergraduates

Measurements on the wear on ordinary and Polyglas tires.

Acceleration of Volkswagens.

Duration time for execution of a freshman program on an IBM 360 computer.

Variations in precision and accuracy of a number of timepieces.

Deviation of resistor values from rated value.

The variability of the temperature in a bulk curing system.

Uniformity of resilient flooring product.

Measurement of transistor H parameters.

Measurement of time for hot tap water to reach 80°C.

Voltage variation in C cells.

Life characteristics of different brands of D cells under periodic and continuous load.

Measurement of the amount of heat emitted by light bulbs of various wattages.

Tire air pressure variations compared to specified values.

Determination of the number of kilowatt hours of electricity used by the average person per week.

Measurement of temperature versus river depth.

Relationship between density and moisture content of various types of wood.

Measurement of the wearing away of a bar of soap.

Percentage of carbohydrates in the sandwich-style loaf of bread.

Lifetime of the average 6- and 2.4-v flashlight bulb.

Fuel consumption curve of an automobile.

Measurement of solids in motor oil.

Measurement of the quantity of rubber deposited on aircraft runways.

Roughness and frictional interaction between various surfaces.

Measurement of the wear pattern of ignition points.

Thickness variation in textbook paper.

Measurement of errors in chaining.

Coefficient of linear expansion of structural reinforcing rod.

Determination of the contents and density of car exhaust fumes.

Photo variations for various f stops and time settings.

Deterioration rate of asphalt with respect to time in use.

Variation in measuring methods applied to the heights of buildings.

Reflectance of light from a mirror.

Measurement of the depth, width, and temperature of a creek.

Measurement of the speed variation of cars on university roads.

Variation in tire traction for different types of tread.

Measurement of the variation in conduction with temperature for a *PN* junction.

Variation in the amount of solids suspended in a creek.

Average height of Canadian population.

Strength of graphite.

Transistor parameters.

Effect of age and environment on the frequency range of the human ear.

Variations in room temperature with variation of outside conditions and heating system.

Measurement of relative humidity in domestic rooms.

Measurements of the breaking strength of automobile safety glass.

Measurements of the filament resistance in light bulbs.

Measurement of the strength of facial tissues.

Measurement of the strength of various types of paper.

Efficiency of a prototype electric motor.

Relationship between film speed and grain size.

Variation in specific gravity of the human body.

Measurements of the traffic flow through an intersection.

Study of the bounce of a golf ball as a function of the compression strength.

Calculation of the temperature of electrical resistance per degree centigrade of iron and copper samples.

Loss of weight and volume of side-bacon when cooked.

Variability and efficiency of home furnace thermostats.

Accuracy of automobile odometers.

Settlement of cereals in packages.

Holding strength of thumbtacks.

Braking distance variation at intersections.

Volume of eggs in specific grade size.

Flight behavior of paper airplanes.

Revolutions per minute of turntables.

Thermal characteristics of a study lamp.
Tar content of cigar versus pipe.
Study of utility of notice boards.
Effect of light on the growth rate of plants.
Variation of temperature with height above the ground.
Weight variation in construction materials due to erosion.
Water content of corn kernels.
Patterns of preference (left hand versus right hand).
Acceleration of hockey players.
Efficiency of cigarette filters.
Effectiveness of alarm in alarm clock.
Time sequence of traffic signals.
Wet strength of facial tissue.
Percentage of bent tin cans in a supermarket.
Temperature distribution in a house.
Strength of elastic bands.
Water content in fruit and vegetables.
Reading speeds of students.
Ratio of student height to maximum arm span.
Change in temperature in an automatic dryer as drying time increases.
Brittleness of bricks.
Examination of the angle of slant of handwriting.
Axial loading of drinking straws to failure.
Deformability of springs.
Comparison of water absorbancy potentials of paper fibers and cloth fibers.
Measurement of the linear coefficient of expansion (steel).
Deflection of small cantilever beams.
Measurement of force and acceleration of a mass that moves on an inclined plane.
Determination of the accuracy and factors affecting the accuracy of triangulation.
Relative filament temperatures of incandescent tungsten lamps in the base-up and base-down positions.
Determination of g using various lengths of pendulum.
Measurement of the degree of curvature of a railway track.
Relationship between sound intensity and distance.
Comparison of the muscular reaction times of people.
Variation of light during sunset.
Average amount of slide rule error by freshman engineers.
Fracture energy of glass.
Measurement of the heating efficiency of a stove top element.
Consistency of mass-produced wood screws.
Average walking speed of people.
Average strength of a strand of hair.
Average amount of smokable tobacco and dollars discarded per package of cigarettes.
Physical properties of a glass cantilever beam.
Measurement of the change in Q and inductance of an inductor as an iron ferrite core is inserted.
Strength of a supermarket paper bag.
Determination and examination of the time constant of two different thermometers.
Measurement of the velocity and penetration power of a projectile fired from a crossbow.

Index

Index

DATE DUE

DATE DUE			
FEB 1 0 1981			
FEB 2 4 1981			
FEB 2 4 1981			
MAR 2 4 1981			
APR 7 1981			
FEB 1 '89			
MAR 3 0 1998			
APR 2 0 2001			
MAY 0 4 2001			
AUG 1 1 2015			
SEP 1 4 2015			
GAYLORD			PRINTED IN U.S.A.